IN MINUTES

GILES SPARROW

PHYSICS IN MINUTES

GILES SPARROW

CONSULTANT: PROFESSOR DAVID W. HUGHES

Quercus

CONTENTS

Introduction

There's an old classroom joke that neatly defines the three major school sciences thus: 'If it wriggles, it's biology. If it smells, it's chemistry. And if it doesn't work, it's physics.' And sadly, while physics is on the receiving end of this particular punch line, those definitions will probably ring a bell with a lot of people. I certainly served my classroom time fudging the results of experiments to measure acceleration due to gravity or demonstrate Boyle's law.

But the classroom experience does physics a disservice – in reality it is perhaps the oldest and certainly the deepest of all the sciences, revealing fundamental truths about the way in which the Universe operates. Although it has taken centuries for us to realize it, physical principles underpin many of the other sciences and all of modern technology. Some would even argue that astronomy, chemistry, and even molecular biology are, in essence, 'applied physics'.

As a science, physics can be both beguilingly simple and bewilderingly complex. The simple equation of motion and laws of electricity and magnetism we learn (and all too soon forget) in the classroom are the same ones that are ultimately applied to building awesomely sophisticated machines such as the Large Hadron Collider, while the alluring handful of fundamental particles that seem to explain the nature of matter itself have properties that can only be explained by theories calling for half a dozen or more additional, unseen dimensions. Even a 'thought experiment' as simple as putting a cat in a box with some knock-out gas turns out to have far-reaching implications for the nature of reality itself.

This book takes you on a journey through the whole of physics, from vaguely remembered classroom experiments to the cutting edge where science blurs into philosophy. With 200 topics and an avowed intent to avoid daunting equations wherever possible, we can barely scratch the surface, but by the time you reach the end, I hope you'll find your memories stirred and your thoughts provoked. Above all else, I hope you'll be convinced that, contrary to that old joke, *physics works*.

Classical mechanics

Mechanics is the field of physics that concerns itself with the behaviour of objects in motion or subject to forces. With origins dating back to ancient Greece, it is the oldest area of physics, and is referred to as 'classical' because it involves laws that were understood long before the twin 20th-century scientific breakthroughs of quantum theory and relativity. These laws can still be applied to explain most common phenomena in the Universe that occur from the scale of atoms and molecules upwards. It's only in extreme situations involving high speeds, strong gravitational fields or very small scales that the more recent breakthroughs offer a more accurate description of what's really going on.

Classical mechanics describes aspects of the Universe that range from simple machines to the orbits of planets, and the properties of solids, liquids and gases from the atomic level up to everyday or 'macroscopic' scales.

Underlying all these different phenomena is an elegantly simple model of colliding bodies, and simple laws that govern their interaction and behaviour. The techniques of classical mechanics allow us to calculate the gain, loss and transfer of energy between particles, and predict their behaviour in a variety of situations.

The most important of these laws are Newton's laws of motion and gravitation, described by the English physicist Sir Isaac Newton in his 1687 book on the *Principles of Natural Philosophy* (best known from its Latin title as simply *The Principia*). Newton's laws describe the motion of objects under the influence of forces (see page 24), and the strength of attractive gravitational forces between large masses (see page 32). As such, they provide a complete description of many natural phenomena that only breaks down in extreme conditions – on the submicroscopic scales of quantum physics or in the situations of extreme speed or gravity described by relativity. Indeed, Newton's laws do such a good job of describing the everyday world around us that Newtonian physics is often used as a synonym for the entire field of classical mechanics.

Speed, velocity and acceleration

The concepts of speed, velocity and acceleration are vital to describing the motion of all bodies in mechanics. Speed and velocity are commonly treated as if they are interchangeable, but strictly speaking, speed is a measure of a body's rate of motion in *any* direction, while velocity is a measure of motion in a *specific* direction. Both can be measured in the same units (such as metres per second), but speed is a directionless or 'scalar' quantity, while velocity is a directed or 'vector' quantity. In most situations, it's far more useful to know an object's velocity than its speed.

Acceleration is another vector quantity – it measures the rate of change in an object's velocity (in units such as metres per second per second, often written m/s^2). Acceleration occurs only when an external force is applied to the object, and depending on the direction of that force, can result in a reduction in its velocity (sometimes termed deceleration) or a change in direction as well as magnitude of its velocity.

When a car accelerates at a constant rate (a), its velocity (v) increases steadily, while the total distance travelled (x) increases exponentially.

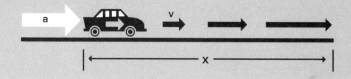

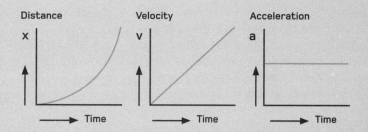

Distance

x

→ Time

Velocity

v

→ Time

Acceleration

a

→ Time

Mass, inertia and weight

The related quantities of mass, inertia and weight describe an object's innate susceptibility to external forces of acceleration. Mass and weight are often (and incorrectly) used interchangeably – an object's mass is directly related to the amount of material it contains, and only changes if material is added or removed (or in extreme situations described by Einstein – see page 374). Inertia is an object's tendency to resist any change in its motion – it is directly proportional to the mass an object contains, and is not usually measured separately, despite being an important concept in Newtonian physics.

Weight, meanwhile, is a measure of the *force* experienced by a given mass within a gravitational field. Since it is a force, it is properly measured in units called newtons, where 1 newton is the force required to accelerate a 1-kilogram (2.2-lb) mass at a rate of 1 metre per second per second (40 in/s^2). Since the acceleration at Earth's surface due to gravity is 9.81 m/s^2 (32.2 ft/s^2) the weight of a 1-kilogram mass is 9.81 newtons.

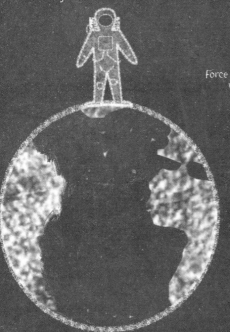

Mass = 120 kg
Force of gravity on Earth = 9.81 N/kg
Weight = 120 x 9.81 = 1177 N

Mass = 120 kg
Force of gravity on Moon = 1.62 N/kg
Weight = 120 x 1.62 = 194 N

Friction

In almost any real-world mechanical 'system', moving objects experience a force of drag that tends to slow them down, while static objects experience a force that tends to prevent motion. These forces, created by interaction with an object's surroundings, are collectively known as friction. Depending on the nature of the system, friction can take various forms. Dry friction occurs between solid surfaces, and may be either kinetic (if the surfaces are in motion) or static. Fluid friction occurs between layers in a liquid or gas, while internal friction resists forces attempting to deform solid bodies.

Three essential laws govern dry friction: Amontons' firstlaw states that frictional force is proportional to the applied load (the proportion of the object's weight pressing down on the surface), while his second law states that it is independent of the areas in contact. Coulomb's law of friction, meanwhile, states that kinetic friction is also independent of the 'sliding velocity' between two surfaces.

The amount of dry friction experienced by an object sitting on a surface can be defined by a coefficient of friction, μ, that depends on the nature of both materials involved.

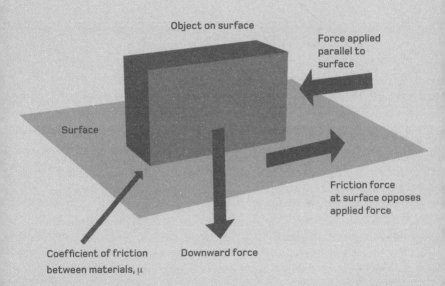

Object on surface

Force applied parallel to surface

Surface

Friction force at surface opposes applied force

Coefficient of friction between materials, μ

Downward force

Momentum

An object's momentum is a property that may be best described as its 'mass in motion' – it is calculated by multiplying a body's mass by its velocity, and is measured in units such as 'kilogram metres per second' (kg m/s). Because it is derived from velocity rather than simply speed, momentum is a vector quantity, with a direction as well as a magnitude. In addition to this linear momentum, a rotating body may also have angular momentum (see page 34).

Linear momentum changes due to the application of a force: the change in velocity is proportional to the change in velocity (so accelerating an object to twice its original velocity also doubles its momentum). Importantly, momentum is a 'conserved' quantity – in a closed system of colliding bodies subject to no external forces, the total momentum always remains the same. Collisions between pool balls on a table or beads suspended in a 'Newton's Cradle' offer clear demonstrations of this principle at work.

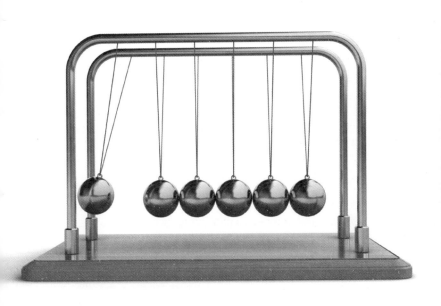

Work, energy and power

Three fundamental concepts can help to understand the behaviour of mechanical systems. In physical terms, work is the application of force to a body possessing mass, which then moves – a force is said to 'do work' when it moves an object in a particular direction. It is calculated by the simple formula

$$W = f \times d$$

where f is the force applied to the object and d the distance moved. Force is measured in newton metres or joules. For example, when a force of 10 newtons is applied to move a body for 12 metres, 120 joules of work has been done.

Energy is a slightly more nebulous but even more important concept – it can be described as the capacity for doing work, inherent to an object or physical system. Energy, like work, is measured in joules, and it can neither be created nor destroyed, but is always conserved throughout a given system (on scales up to and including the Universe as a whole). However, energy can take multiple forms, from relatively 'useful' ones such as kinetic and potential energy (see page 20) to more diffuse and inaccessible forms such as ambient heat and sound.

In this way, a system can lose the capacity to do useful work over time, which is why, for example, a rubber ball does not keep bouncing forever. An entire field of physics, known as thermodynamics, is dedictated to understanding and modelling these concepts (see page 104-5).

Finally, power is simply the rate at which work is done or energy is used. It is measured in watts, where 1 watt is equivalent to 1 joule of work done (or energy expended) per second. To extend the previous example, if a force of 10 newtons takes 5 seconds to move an object over a distance of 12 metres, the power expended to do the work is 120/5 = 24 watts.

Kinetic and potential energy

An object's kinetic energy is the energy it possesses due to its motion, and can be calculated from the amount of work required to accelerate the body from rest to its current speed (see page 18). While this can be calculated from first principles for any object, in classical situations where relativity is not a factor, it is easier to use the simple formula $K.E. = \frac{1}{2} mv^2$, where m is the object's mass and v its velocity.

Potential energy is the energy an object has thanks to its position in a system – equivalent to its capacity to do work. Its most familiar form is gravitational potential energy, calculated by the formula $P.E. = mgh$, where h is the object's height and g the force of gravity (equivalent to 9.81 newtons per kilogram).

Other forms of potential energy include elastic, electrical, chemical and magnetic, and like all forms of energy, they can be exchanged from one form to another. In fact, all physical processes can ultimately be defined as transfers of energy.

Initial
velocity
= 0

Mass m

For an object suspended at
height Δh above ground:

Kinetic energy = 0
Gravitational potential energy = $mg\Delta h$

Acceleration
due to
gravity = g

Height of
object = Δh

Allowed to drop the object's
potential energy is converted to
kinetic energy of movement as it
reaches velocity v:

Kinetic energy = $\frac{1}{2}mv^2$
Potential energy = 0

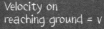

Velocity on
reaching ground = v

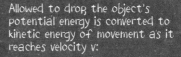

Elastic and inelastic collisions

Physicists recognize two distinct categories of collision between objects. An elastic collision is one in which both the momentum and the kinetic energy (K.E.) of a system are conserved (in other words, a measure of these quantities after the collision will produce the same result as a measurement before the collision). An inelastic collision, in contrast, conserves momentum, but not K.E. However, the overall energy of a closed system is always conserved, so any change in K.E. must be accompanied by a transfer of energy to or from other forms.

Most collisions are inelastic: for example, when two pool balls collide, some K.E. is transformed into heat (vibrations of the atoms within the balls) and sound. In fact, the only truly elastic collisions are those between atoms themselves, but fortunately in practice many systems can be treated as elastic – either because the change in K.E. is negligible, or because losses and gains in K.E. can be balanced statistically (see page 48).

In elastic collisions, both momentum and kinetic energy are conserved:

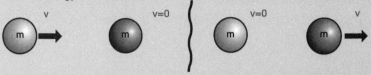

Here, the kinetic energy is transferred completely to the second mass. The overall momentum in the system remains mv throughout, while the K.E. remains $\frac{1}{2}mv^2$.

In an inelastic collision, momentum is conserved, but kinetic energy is not:

Here, the two masses stick together. In order to preserve an overall momentum of mv, their combined velocity must now be $v/2$, so the K.E. of the system is halved.

Newton's laws of motion

In his 1867 *Principia*, Isaac Newton laid down three laws of motion that are usually given in the following form:

1) An object will continue in uniform motion (or remain at rest) unless acted upon by an external force.
2) An object's acceleration (a) is in the direction of, and directly proportional to the sum of the forces (F) acting on it, and inversely proportional to its mass (m). This can be written down in the equation $a = F/m$ (often recast in the form $F = ma$).
3) The force generated by one body on another is balanced by an equal and opposite force exerted by the second body on the first: 'every action has an equal and opposite reaction'.

These laws explain the mechanical behaviour of both microscopic and macroscopic (large-scale) systems, and form the theoretical bedrock of classical physics.

1 Force F applied by finger onto marble of mass m
2 Acceleration of marble a = F/m
3 Reaction force (pressure) F
   from marble to finger

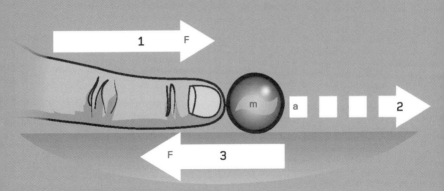

Equations and graphs of motion

A variety of mathematical equations can be used to describe the properties of bodies in motion, but perhaps the most familiar and useful are five that apply to situations involving constant acceleration. Sometimes called the SUVAT equations, they link the variables s (displacement or distance travelled), u (initial velocity), v (final velocity), a (acceleration) and t (time). The equations can be briefly stated as follows:

$$v = u + at \qquad\qquad s = ut + \tfrac{1}{2}at^2$$

$$s = \tfrac{1}{2}(u+v)\,t \qquad\qquad v^2 = u^2 + 2as$$

$$s = vt - \tfrac{1}{2}at^2$$

Another way of looking at the situations involving constant acceleration is to draw a graph comparing velocity and time, as shown opposite. The SUVAT equations and the geometry of the graph both reveal the same relationships.

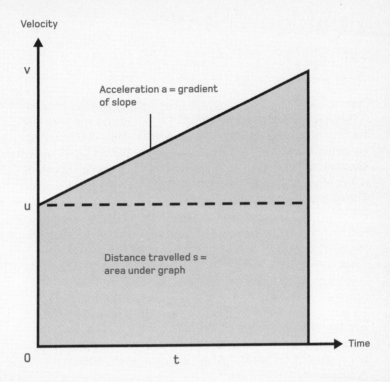

Velocity

v

Acceleration a = gradient of slope

u

Distance travelled s = area under graph

0   t   Time

Orbits

An orbit is the path that one body follows around another under the influence of an attractive force between the two. According to Newton's first law of motion, objects should continue in straight-line motion unless subject to an external force, so it's the role of the attractive force to pull one body into a looping path or orbit. At any point in space, the motion of a body in orbit around another can be considered as two components – a radial element along the straight line between the bodies, and a lateral element at right angles to it, pointing 'along' the orbit. Stable orbits arise where the attractive force pulling on the orbiting body exactly balances the radial component of its motion, resulting in only lateral motion. The precise direction of the lateral motion is different at each point in space, resulting in a curving path.

Although the simplest orbit is a perfect circle, Kepler's laws (see page 30) show how closed orbits actually follow stretched or elliptical paths, of which the circle is a special case.

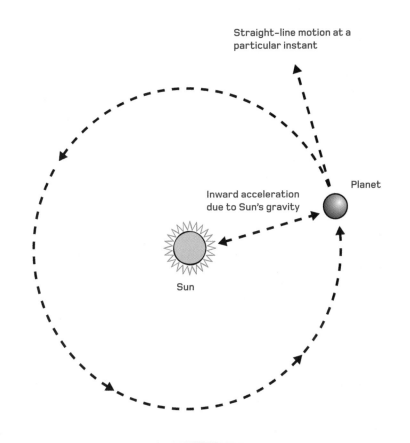

Straight-line motion at a
particular instant

Planet

Inward acceleration
due to Sun's gravity

Sun

Kepler's laws of planetary motion

In 1609, German mathematician Johannes Kepler published a new description of the cosmos. His careful analysis of planetary motions led him to realize that their orbits around the Sun obey three simple laws:

1) Each planet's orbit is an ellipse with the Sun at one focus.
2) A line between planet and Sun sweeps out equal areas in equal time (so planets move faster when closer to the Sun).
3) The square of each planet's orbital period is proportional to the cube of its semi-major axis.

The idea that the planets orbit the Sun had been gaining in popularity since Nicolaus Copernicus published his work on the subject in 1543, but astronomers were still wedded to the philosophical 'perfection' of circular orbits. By breaking free of this concept, Kepler revealed the true motion of bodies under the influence of gravity – pointing the way for Newton's more generalized laws of motion and gravitation.

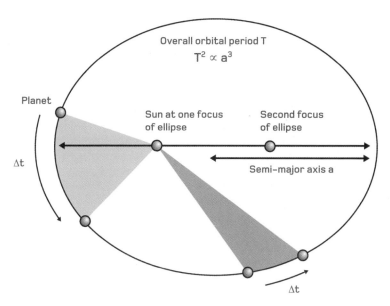

Orbit sweeps out equal areas
in equal times Δt

Newtonian gravity

Gravitation – the force that attracts bodies with mass towards each other – is a central concept for all of physics, and is particularly essential to mechanics. In 1589, Italian scientist Galileo Galilei showed that Earth's gravity affects objects in the same way regardless of their mass – if factors such as air resistance can be removed, all objects experience the same force pulling them towards the ground.

A century later, the sight of a falling apple supposedly inspired Isaac Newton to a great conceptual leap: what if the force drawing the apple towards the ground was universal, and the same force that keeps the Moon in orbit around the Earth? Building on Kepler's laws of planetary motion (see page 30), Newton was able to develop his own generalized laws of motion, and then a law of universal gravitation. This states that the attractive force between two objects is proportional to the product of their masses multiplied together, and inversely proportional to the square of the distance between them.

1 Acceleration due to gravity
  = 9.8 m/s (32.2 ft/s) per second

2 Force of gravity on a 1 kg (2.2 lb)
  object = 9.8 newtons

Angular momentum

In contrast to the linear momentum of objects with mass moving in straight lines, angular momentum (L) is exhibited by rotating objects. It is a vector quantity with both magnitude and direction, and is always measured relative to a particular axis, the system's axis of rotation. It can be calculated using the equation $L = I\omega$, where I is the body's 'moment of inertia', a measure of its resistance to rotation, while ω is its angular velocity (revolutions per unit time), the speed with which its angular direction changes relative to the system's axis.

Just like linear momentum, angular momentum is conserved in a closed system unless an external rotational force, known as a torque, is applied. Conservation of angular momentum helps explain many natural phenomena, from the way in which a pirouetting ice skater spins faster by drawing his or her arms inwards, to the way that stars and planets begin to rotate more rapidly as they form out of whirling clouds of gas and dust.

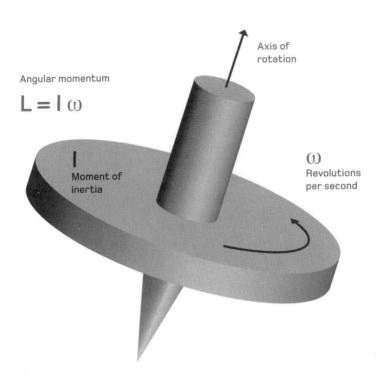

Axis of
rotation

Angular momentum

$$L = I\omega$$

I
Moment of
inertia

ω
Revolutions
per second

Centripetal and centrifugal forces

A centripetal force is any force that makes a body follow a curved path – it always acts at right angles to the object's motion, towards a point known as the instantaneous centre of curvature. The force may be provided by gravity (as in the case of an orbit), or by tension along a cord or bar, for example, when an object such as a ball is whirled around an athlete's head. The forces involved can be calculated from the equation $F = ma = mv^2/r$, where m is the object's mass, a is the accelerating force, v is the velocity along a tangent at any particular instant and r is the radius of the curve. It can also be expressed as $F = mr\omega^2$, where ω is the object's angular velocity.

In these types of situation, 'centrifugal force' is an apparent force that pushes or pulls a body away from the centre of rotation. It may be a real force produced in reaction to the centripetal force, or an entirely illusory or 'fictitious' force (see page 38) resulting from a change of reference frames.

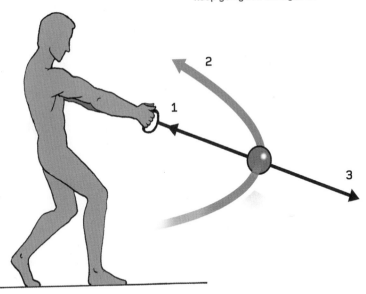

1 Inward force of tension between ball and athlete

2 Curved path of ball around athlete

3 Fictitious 'centrifugal' force is caused by the ball's tendency to keep going in a straight line

Coriolis effect

The Coriolis effect is most familiar from its effect on the rotation of Earth's water and weather systems, but it is in fact a more general effect that neatly demonstrates how 'fictitious forces', such as centrifugal force, can arise. The effect is an apparent deflection to the motion of an object moving in a straight line when it is viewed from within the same rotating 'frame of reference'. If the frame of reference is spinning clockwise, the deflection is to the left, while if it is anticlockwise, the deflection is to the right.

Earth's relatively slow rotation means that the Coriolis effect only makes itself noticeable over relatively large distances and long periods of time – contrary to popular folklore, it could only affect water draining down a plughole in ideal circumstances. Larger masses of air or water subject to no other forces are driven to form 'inertial circles' whose radius increases closer to the equator, and which rotate clockwise in the northern hemisphere and counterclockwise south of the equator.

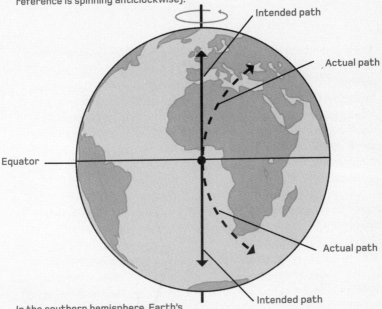

Driven by Earth's rotation, the Coriolis effect deflects paths to the right in the northern hemisphere (where the frame of reference is spinning anticlockwise).

Intended path

Actual path

Equator

Actual path

Intended path

In the southern hemisphere, Earth's rotation is perceived as 'clockwise', and Coriolis deflections are to the left.

Simple machines

In mechanical terms, a simple machine is a device that offers some sort of 'mechanical advantage' – usually, this means a system that offers a way of reducing the force that must be applied to move a particular object. Simple machines take advantage of the fact that work done in moving an object is given by multiplying the force by the distance moved (see page 18). Therefore it's possible to do the same amount of work by, for example, applying half the force over twice the distance.

The lever demonstrates this most clearly – when the fixed pivot-point or fulcrum is located one-third of the way along its length, so that the point at which the force is to be applied is twice as far from the fulcrum as the object to be lifted, it allows half the force applied over twice the distance to do the same work. Other simple machines include the inclined plane, the wedge, the wheel and axis, the pulley and the screw – all of which reduce the force applied to do a job, in exchange for extending the time or distance over which the force is required.

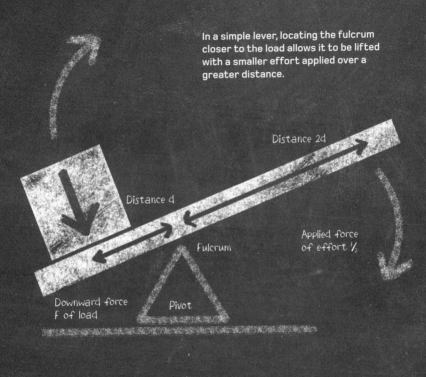

In a simple lever, locating the fulcrum closer to the load allows it to be lifted with a smaller effort applied over a greater distance.

Distance 2d

Distance d

Fulcrum

Applied force of effort $\frac{F}{2}$

Downward force F of load

Pivot

Deformation

In physical terms, deformation is a change in the shape of a solid object, due to either the application of a force or the heating of the object. Deformation under stretching forces is perhaps the most familiar form of the phenomenon – thanks to their internal atomic or molecular structures, solid materials resist forces of compression quite well (although they may still deform when subjected to twisting or shearing forces). Physicists distinguish between situations where deformation permanently alters a material's shape (plastic deformation) and cases where an object can recover its original form once the deforming force is removed (elastic deformation). Excess strain will cause an object to 'fail' or break.

Hooke's law states that, for elastic deformation, the extent of deformation X (for example, in the stretching of a spring) is proportional to the deforming force F applied to it. In equation form: $F \propto X$ or $F = kX$, where k is a constant measuring the object's resistance to deformation known as its stiffness.

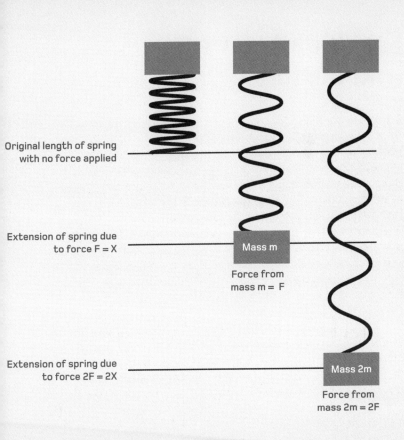

Original length of spring
with no force applied

Extension of spring due
to force F = X

Mass m

Force from
mass m = F

Extension of spring due
to force 2F = 2X

Mass 2m

Force from
mass 2m = 2F

Elasticity, stress and strain

When an object is subjected to a deforming force, its response is governed by Hooke's law (see page 42) and a constant, known as its stiffness, that depends on both the material from which the object is made and its shape. Another constant, known as elasticity, allows us to assess the properties of the raw material itself. It is usually described by the simple equation λ = stress/strain, where λ is the elastic modulus (a measure of elasticity), stress is the force applied per unit area of the material (measured in pascals) and strain is the material's change in length, divided by its original length, provided the material's 'elastic limit' is not exceeded.

In fact, physicists and engineers use several elastic moduli to describe a material's properties in different situations. Young's modulus describes a material's elasticity when subjected to forces along a single axis, shear modulus describes a material's overall tendency to hold its shape, and bulk modulus describes its resistance to compression from all sides.

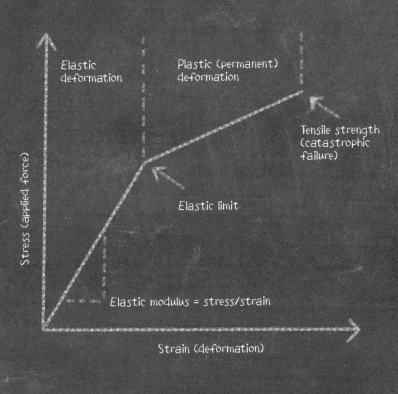

Elastic deformation

Plastic (permanent) deformation

Tensile strength (catastrophic failure)

Elastic limit

Stress (applied force)

Elastic modulus = stress/strain

Strain (deformation)

Gas laws

The behaviour of gases is an important area where the laws of mechanics can be applied statistically to a huge number of particles (individual atoms or, more commonly, molecules containing two or more atoms) in a system. These techniques help explain the 'gas laws' discovered by experimental scientists between the 17th and 19th centuries. Each law describes the behaviour of a sealed system containing a fixed quantity of gas.

Boyle's law states that within a closed system, the volume of a fixed mass of gas is inversely proportional to the pressure it exerts on its container, provided its temperature remains constant. In other words, if you compress a fixed amount of gas (reducing its volume, for example, with a piston), its pressure increases. In equation form, this can be written as $P \propto 1/V$.

Charles' law, meanwhile, states that if the pressure of gas in a system remains constant, then the volume it occupies is directly proportional to its absolute temperature (see page 108). It can be mathematically stated in the form $V \propto T$.

Finally, Gay-Lussac's law states that, for a gas contained in a fixed volume, the pressure exerted on the container is proportional to the absolute temperature: $P \propto T$.

These three laws define a triangular relationship of interdependence, and can be easily combined to show that the product of the gas's pressure and volume is proportional to its absolute temperature:

$$PV \propto T, \text{ or } PV = NkT$$

(where N is the number of molecules of gas in the system and k is a constant known as Boltzmann's constant).

It's important to note that, strictly speaking, the gas laws only describe 'ideal' gases – those composed of point-like particles in random motion that do not interact with each other (see page 48). Fortunately, many real gases behave like ideal gases in all but the most extreme conditions – the exceptions tend to be heavy gases such as refrigerants, or those with strong forces between their molecules, such as water vapour.

Kinetic theory of gases

The kinetic theory is a way of understanding ideal gases through a statistical model of the individual molecules within them. Its development was one of the first successful attempts to apply classical mechanics on a microscopic scale, and it played an important role in proving that everyday materials are composed of invisibly small molecules and atoms.

According to the theory, an ideal gas contains numerous small particles in constant, random motion. Pressure exerted by a gas is the result of collisions between individual particles and the walls of its container, while the gas's temperature reflects the speed with which the particles move. The model assumes that the particles are small compared to the average space between them. For a gas composed of individual atoms, it's therefore easy to show how changes to either pressure (P), temperature (T) or volume (V) affect the other properties, resulting in the ideal gas law. For a system containing n moles of an ideal gas, $PV = nRT$, where R is the 'universal gas constant'.

1 Random motions of gas molecules in container

2 Speed of molecules increases with temperature

3 Collisions of gas molecules with container walls and other molecules exert pressure

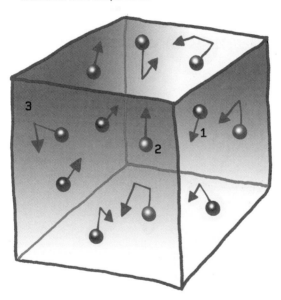

Avogadro's law

_____

A vogadro's law links the volume of a gas to the number of atoms or molecules present. These numbers are measured in 'moles', where 1 mole is equivalent to 6.02×10^{23} particles. This number, known as the Avogadro constant, is the number of atoms of a particular atomic mass m (see page 158) that constitute a mass equal to m grams.

According to the law, at a fixed temperature and pressure, the volume V of an ideal gas will be proportional to the number of moles of gas n that it contains:

$$V_1/n_1 = V_2/n_2$$

Experiments have proved that a single mole of any ideal gas occupies a 'molar volume' of 22.414 litres (4.9 gal) at 0°C (32°F) under an atmospheric pressure of 1 bar, so we can say, for example, that 64 grams of O_2 (molecular oxygen with a molecular mass of 32) will occupy 44.818 litres in these conditions. This offers a convenient means of measuring the mass of gases.

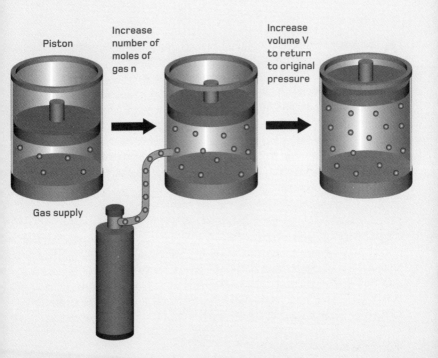

Piston

Increase number of moles of gas n

Increase volume V to return to original pressure

Gas supply

Fluid mechanics

A complex field of science in its own right, fluid mechanics is the branch of physics that concerns itself with the properties of fluids. It attempts to model the often complex processes associated with fluid behaviour, and applies to both liquids and gases. Significantly – and unlike the gas laws – its laws ignore the fact that fluids are made up of large numbers of individual atoms or molecules, in favour of a 'continuum hypothesis'. This assumes the fluid is a single structure and that changes from one part to another are continuous.

In fact, fluid mechanics often makes other assumptions to simplify its models. These include the idea that fluids are incompressible (largely true for liquids but not for gases), and that they may be non-viscous or 'ideal' – in other words, they have no internal friction-like forces to prevent them deforming in response to external forces known as shear stresses. In practice, only superfluids (see page 320) are truly non-viscous, but ideal fluids are still a useful theoretical model.

A fluid's viscosity is analogous to the elastic modulus of solid materials (see page 44) – it measures the ease with which a material flows and deforms. Fluid deformation or shear strain (the change in the fluid's dimensions compared to its original size) changes rapidly, so it's more useful to measure the rate at which the shear strain changes. In equation form:

$$\text{viscosity} = \frac{\text{shear stress}}{\text{rate of change in shear strain}}$$

The resulting coefficient of viscosity is measured in pascal seconds, and the viscosity of water at room temperature happens to have the neat value of 1 millipascal second. Informally, substances with higher coefficients than water are termed viscous, while those with lower ones are non-viscous.

These concepts can also help to identify a useful division among fluids. In normal 'Newtonian' fluids, stress is proportional to the rate of strain at every point, and a constant, predictable coefficient of viscosity can be calculated. In non-Newtonian fluids, viscosity is not constant and can be affected by other factors such as time and the rate of change in stress. Tomato ketchup and paint are both non-Newtonian fluids.

Archimedes' principle

The best-known law of fluid mechanics – and perhaps the most famous physical law of all, on account of the colourful stories surrounding its discovery in the third century BC – states that when a body is immersed in a fluid, the upward force (buoyancy) it experiences is equal to the weight of fluid it displaces. Note that the force depends only on the amount of a fluid that is actually displaced – if the object is light enough to remain afloat, then it is only displacing fluid equivalent to its submerged part, and only that much force will be acting on it.

Archimedes supposedly discovered the principle when he was set the problem of testing if a golden crown made for the King of Syracuse had been adulterated with silver. Although this story does not appear in any of the philosopher's own writings, the principle does allow density to be measured by the formula:

$$\frac{\text{density of object}}{\text{density of fluid}} = \frac{\text{weight}}{\text{weight of displaced fluid}}$$

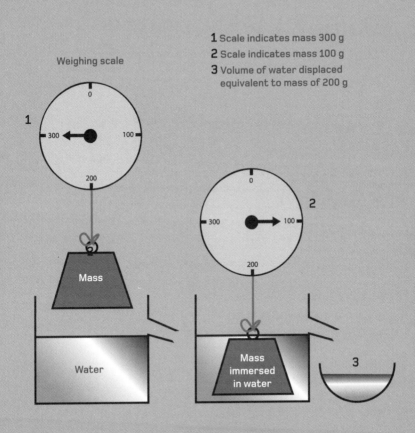

Weighing scale

1 Scale indicates mass 300 g
2 Scale indicates mass 100 g
3 Volume of water displaced equivalent to mass of 200 g

1

Mass

Water

2

Mass immersed in water

3

Bernoulli's principle

One of the most important principles in fluid dynamics seems at first to be counterintuitive, but even so, it is one without which many forms of modern technology could not work. In his 1738 work *Hydrodynamica*, Swiss scientist Daniel Bernoulli explained how, for an ideal fluid, any increase in the speed of flow must be matched by a drop in the pressure or potential energy. If the fluid is flowing horizontally, then it will exert more pressure when it is moving slowly and less pressure when it is moving quickly.

A full application to real-life situations is of course rather more complicated than this, but the principle is useful in a variety of applications. For example, Venturi meters measure the rate of fluid flow by the pressure change caused when the fluid moves between two different gauges of pipe. Aircraft wings, meanwhile, force air to travel at higher speeds across their upper surfaces, creating a pressure difference and an upward-lifting force.

Bernoulli effect on an aircraft wing

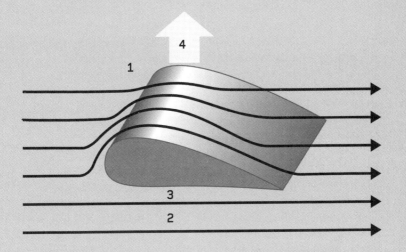

1 Faster flow of air diverted over top of wing
2 Slower flow of air passing below wing
3 Volume of higher pressure
4 Volume of lower pressure creates lift

Chaos

The complex mathematics of chaos theory is a huge and growing field with implications for disciplines including biology, economics and even philosophy. However, from a physicist's point of view they are best summed up by the simple phrase 'sensitivity to initial conditions'. In other words, even in the apparently predictable systems of classical physics, the slightest difference in initial conditions can lead (in certain circumstances) to vastly different outcomes.

US mathematician and meteorologist Edward Lorenz, who invented the famous concept of the 'butterfly effect', neatly summed up the implications when he defined chaos as a situation in which 'the present determines the future, but the approximate present does not approximately determine the future'. In other words, without the ability to measure and control every factor with infinite accuracy, certain systems are unpredictable by their very nature.

The Lorenz attractor depicted in this computer graphic is an equation whose solutions vary chaotically under certain conditions.

Waves

Moving disturbances or oscillations that (generally) transfer energy between locations, waves are a widespread and fundamental aspect of the physical Universe. Understanding wave properties and behaviour is one of the key foundations of modern physics, with implications that extend beyond obvious wave phenomena, such as rippling water and vibrating sounds, into fields as diverse as electromagnetism and the very nature of matter itself.

It's hard to reach a precise and all-encompassing definition of waves, but easy to recognize them when you see them. When asked to think of a wave phenomenon, most people will turn first to ripples on water, and indeed these demonstrate many important aspects of the phenomenon in general.

One thing that a consideration of water waves makes instantly clear, is that while a wave is a moving disturbance, it often leaves the material it passes through in essentially the same place – a floating feather will confirm that when a single ripple passes across a pond, individual areas of water (right down to the molecular level) are only briefly disturbed, and usually return to their previous position and state once the wave has passed. The substance that a wave moves through is known as its medium, and so far as we know, only one type of wave (electromagnetic radiation – see page 124) is capable of existing without a medium.

Water ripples are 'transverse' waves – the up-and-down oscillation in the medium created by their passage is at right angles or perpendicular to the wave's direction of travel or propagation. While this is true for many other types of wave (including electromagnetic ones) it is not universal – in 'longitudinal' waves, the oscillation or disturbance is aligned with the direction of propagation. Sound or pressure waves are a good example of this kind of wave – their propagation involves the 'bunching up' or compression of particles along the direction of movement, followed by a relaxation or expansion as the wave passes. Despite this apparently fundamental difference, however, both types of wave share many measurable properties (see page 62) and are governed by many of the same phenomena.

Wave properties

Any wave can be described by four fundamental properties – velocity, wavelength, frequency and amplitude (or intensity). What's more, the first three of these are linked by simple equations so that if only two are known, it's easy to calculate the third.

Velocity is simply the speed with which the wave itself transfers energy in the direction of its motion. Wavelength, meanwhile, is the distance between two successive maxima or minima of the wave disturbance, often indicated in equations by the Greek letter lambda (λ). In transverse waves, the

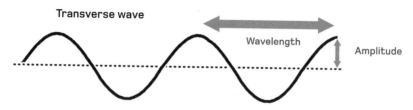

Transverse wave

Wavelength

Amplitude

maxima and minima are defined by the visible peaks and troughs of the overall waveform, while in longitudinal waves, it is defined by the regions of maximum or minimum compression. Despite this difference, the terms 'peak' and 'trough' are used for both types of wave.

Frequency is the rate at which peaks or troughs pass an arbitrary point along the wave, and is measured in cycles per second, or hertz. Finally, amplitude is the wave's 'strength' – the size of the disturbance it generates, measured either by the height of the transverse wave or the degree of compression of a longitudinal one.

A wave's velocity v is easily obtained by multiplying its frequency f by its wavelength λ – in equation form $v = f\lambda$. From this, it follows that $f = v/\lambda$, while $\lambda = v/f$.

Longitudinal wave

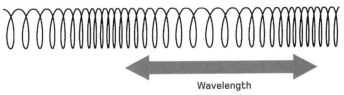

Wavelength

Huygens' principle

First proposed by 17th-century Dutch physicist Christiaan Huygens, Huygens' principle offers a useful means of explaining wave behaviours, such as refraction and diffraction (see pages 82 and 86). Huygens argued that every point along a moving wavefront can be treated as a source of new 'wavelets' spreading out in all directions. As each wavelet encounters obstacles, its behaviour can be predicted through the use of additional wavelets.

This principle neatly explains why waves spread out again after passing through narrow apertures, and also why waves are refracted at the boundary between different media. However, Huygens himself was unable to explain why the wavelets only affect the overall wave on their leading edges. This mystery was only solved by French physicist Augustin-Jean Fresnel in 1816, when he showed that destructive interference between the wavelets (see page 66) would cause their effects to cancel out in all other places.

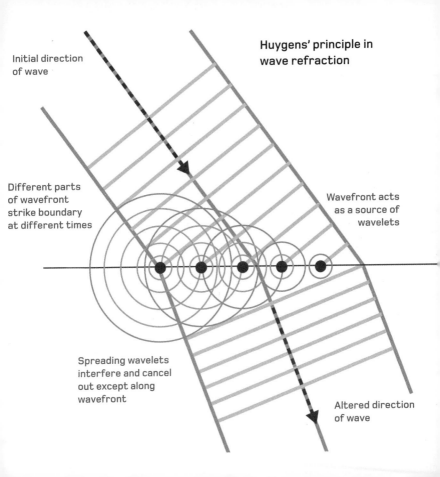

Huygens' principle in wave refraction

Initial direction of wave

Different parts of wavefront strike boundary at different times

Wavefront acts as a source of wavelets

Spreading wavelets interfere and cancel out except along wavefront

Altered direction of wave

Interference

When two waves encounter each other in the same medium, their effects can cancel out or reinforce each other. This effect, known as interference, can be easily demonstrated by throwing two stones into a pond and watching the complex patterns that emerge where their expanding ripples meet.

Wave interference is just one example of a more general physical principle known as superposition: at any point, the amplitude of the disturbance produced by the two combined waves is simply the sum of their individual disturbances. Where two equal 'peaks' meet, the result is a peak of twice normal height, and when equal troughs meet, the result is a trough twice as deep. This is known as constructive interference. Conversely, when a peak and a trough of equal but opposite heights meet each other, they cancel out completely leaving no net disturbance – a phenomenon known as destructive interference.

Perhaps the most familiar form of interference happens when ripples spreading through water overlap with each other.

Modulation

Many of the properties of waves make them ideal for sending information, and indeed since the discovery of radio waves in the 1880s (see page 130), modern civilization has come to rely on them as signal carriers. Today, we can use cheap integrated circuits and computing power to encode and decode information in digital form (see page 238). But traditionally, such signals were created by overlaying or encoding a continuously varying 'analogue' stream of

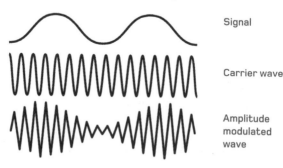

Signal

Carrier wave

Amplitude
modulated
wave

information, such as the signals from a microphone in a distant studio, onto a uniform carrier wave.

In practice, there are two ways of doing this. Amplitude modulation (AM) alters the strength or amplitude of the carrier wave in line with the strength of the signal. The signal is simple to extract. but prone to 'noise' due to random variations in the strength of the wave reaching the receiver. Frequency modulation (FM), meanwhile, shifts the frequency of the carrier wave slightly to either side of a central frequency, representing the original signal at one remove. Extraction is more complex, but because the frequencies of radio waves are less prone to disruption than their amplitudes, the reconstructed signal is much clearer.

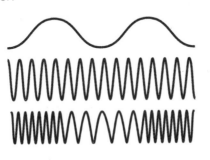

Signal

Carrier wave

Frequency modulated wave

Sound waves

The phenomenon we perceive as sound is a complex pressure wave that is transmitted through media such as air, water and solid objects in the form of longitudinal waves. When these pulses of pressure reach our eardrums they create vibrations that stimulate our sensory nerves and are interpreted as sounds of different pitch depending on their frequency. Sound waves travel at different velocities in differerent media, but the oft-quoted 'speed of sound' is the velocity in air at a standard atmospheric pressure of 1 bar and a temperature of 20°C (68°F) – some 343 metres per second (1,230 kilometres or 767 miles per hour). In contrast, the speed of sound in water of similar temperature is 1,482 metres per second (3,315 mph).

In solids, sound travels as both longitudinal and transverse waves. The latter create side-to-side 'shear stresses' in the material. Sound waves travel in solids at even higher speeds – when they are generated by movements in Earth's crust, they are known as seismic waves and experienced as earthquakes.

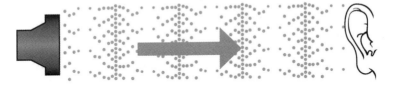

Speaker   Compression of air particles   Ear

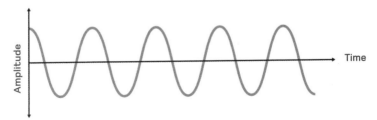

Representation of sound wave reaching ear

Doppler effect

First described by Austrian physicist Christian Doppler in 1842, the Doppler effect is a phenomenon that alters the wavelength and frequency of waves when either the source or the observer is in motion. If the source and observer are getting closer to each other, then the peaks of waves will pass the observer at higher frequency than if they were static, and be measured as having shorter wavelengths. If source and observer are moving apart, then the wave peaks will reach

Wave stretched out when source recedes from observer

the observer more slowly and the wavelength will appear to be longer.

We commonly experience this effect when we hear an emergency siren approach, pass us and then recede, changing from higher to lower pitch (shorter to longer wavelengths) in the process. But the same phenomenon affects waves of all types, and its most important manifestation from a scientific point of view is the Doppler-shifting of electromagnetic radiation. This causes light from objects such as distant stars that are approaching us to be 'blue shifted' to shorter wavelengths, while light from receding objects such as distant galaxies is stretched or 'red shifted'.

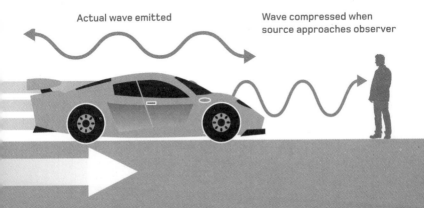

Actual wave emitted

Wave compressed when source approaches observer

Acoustics

Although often thought of as the study of sound, acoustics in fact encompasses the study of all mechanical waves – oscillating disturbances of matter that pass through solids, liquids and gases. Its centrepiece is the acoustic wave equation – a complex mathematical equation that describes the way that acoustic pressure (linked to the motion of individual particles within the medium) changes with location and time.

The study of acoustics is often traced back to the ancient Greek philosopher Pythagoras, who investigated harmonics and resonance as early as the sixth century BC (see page 76). Italian scientist Galileo Galilei and French mathematician Marin Mersenne discovered the properties of vibrating strings around the turn of the 17th century, and Galileo even pinpointed the mechanism behind human sound perception.

However, the modern understanding of acoustics relies on the complex mathematical calculus developed in the late 1600s by Isaac Newton and the German mathematician Gottfried Wilhelm Leibniz. Newton was the first to calculate the speed of sound analytically rather than experimentally, and derived equations for working out the speed of sound (c) in different media. Today, the most generalized form is the Newton-Laplace equation:

$$c = \sqrt{(K/\rho)}$$

where K is the bulk modulus of the medium (see page 44) and ρ is its density.

Another important consideration in the field of acoustics is the way in which sound is perceived. This depends on both the pressure and frequency of sound waves – the average human ear can detect variations in sound pressure level (SPL) down to one-billionth of the ambient pressure, and frequencies between 20 and 20,000 Hertz. Sound pressure is measured in decibels – a logarithmic scale able to accomodate the extreme differences in pressure level that can be experienced, while the spectrum of mechanical wave frequencies is divided into the audible or sonic, the infrasonic (with frequencies below 20 Hz) and the ultrasonic (with frequencies above 20,000 Hz).

Harmonics and resonance

When the medium that carries a wave is finite in extent, it often displays a fundamental frequency. This is the lowest frequency that a repeating or periodic wave can sustain in the medium – for instance in a vibrating violin string it corresponds to a wavelength twice the length of the string.

Harmonics are waves whose frequencies are integer (whole-number) multiples of the fundamental frequency, and sustain themselves in the same way as the fundamental frequency. For a fundamental frequency f with corresponding wavelength λ, the first harmonic has frequency 2f (and a wavelength of $\lambda/2$, the second harmonic has a frequency 3f and wavelength $3\lambda/2$, and so on. Such harmonic systems or 'oscillators' are common in nature, but also common in musical instruments where they produce a complementary range of sound frequencies.

Resonance is a phenomenon in which the amplitude of waves in an oscillator becomes much stronger when it is forced to vibrate at its fundamental frequency or a harmonic. It is most familiar from music, but occurs in everything from engineering structures to electrical circuits and even down to subatomic particles.

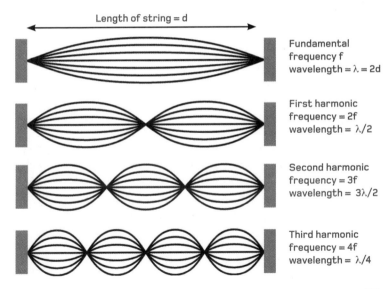

Length of string = d

Fundamental frequency f
wavelength = λ = 2d

First harmonic
frequency = 2f
wavelength = λ/2

Second harmonic
frequency = 3f
wavelength = 3λ/2

Third harmonic
frequency = 4f
wavelength = λ/4

Optics

Light and other closely related radiations form a family of waves with unique properties. While we cannot observe its wave structure directly, it exhibits unmistakeably wave-like behaviour such as diffraction and interference. It is a transverse wave that can pass between different media with ease, and since it is not limited in its orientation, it can exhibit interesting qualities, such as changing polarization (see page 92). Since visible light is one of the few forms of electromagnetic radiation not blocked by Earth's atmosphere, our eyes have evolved to collect it and transform it into nerve signals that can be processed by the brain. As a result, it has become our chief way of interpreting the world around us.

Little wonder, then, that an entire field, known as optics, is devoted to the study and manipulation of light and related radiations. This field encompasses not only the study of light's raw properties, but also ways of harnessing it to create optical instruments that enhance our natural faculties.

Reflection

When a wave encounters the boundary between two different media, it occasionally changes direction in such a way as to bounce back into its original medium – this is the technical definition of reflection.

In general terms, there are two types of reflection: specular and diffuse. In specular reflection, waves rebound in a direction related to their 'angle of incidence' and the orientation of the reflecting surface (see opposite). As a result, many of the characteristics of the incoming wave are preserved, and the reflecting surface creates a 'virtual image', that appears to lie on the other side of the boundary. Specular reflection is most familiar in the case of light, and requires the extremely smooth reflective surface we call a mirror. In diffuse reflection, meanwhile, waves scatter in all directions from the reflecting surface, so while the wave's energy is reflected, the image is lost – an effect seen, for example, on rough cloth and metallic surfaces.

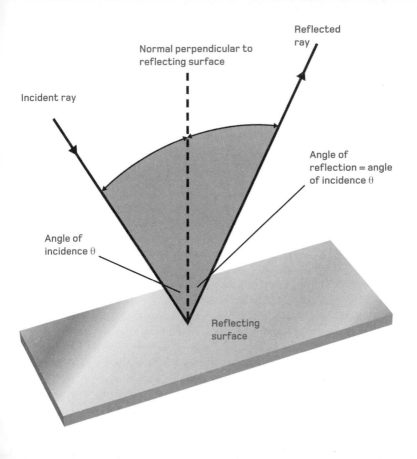

Reflected ray

Normal perpendicular to reflecting surface

Incident ray

Angle of reflection = angle of incidence θ

Angle of incidence θ

Reflecting surface

Refraction

A wave is said to be refracted when it changes direction as it crosses the boundary between two different media. The phenomenon is probably most familiar from the way in which an apparently straight object such as a drinking straw appears to bend as it crosses the boundary between air and water, due to refraction in the path of light from the underwater section.

Refraction is caused when different media transmit a wave at different speeds. In optics, the ratio of the speed of light in a vacuum to the speed in another medium is known as the medium's refractive index. When light travels from a medium with a lower refractive index into one with a higher one, refraction causes it to bend towards the normal – an effect best explained by Huygens' principle (see page 64). Moving from a medium with a higher refractive index to a lower one, the wave is bent in the other direction. Snell's law (opposite) links the angles involved with the refractive indices of the light-transmitting media in one equation.

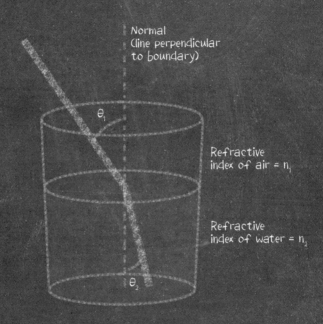

Normal
(line perpendicular
to boundary)

θ_1

Refractive
index of air = n_1

Refractive
index of water = n_2

θ_2

Light rays crossing a boundary into a medium with a higher
refractive are bent towards the normal in accordance with
Snell's law: $n_1\sin\theta_1 = n_2\sin\theta_2$.

Total internal reflection

When a wave encounters the boundary between one medium and another, it normally splits into two elements of less intensity. One part passes successfully through the boundary (usually being refracted in the process), and the rest is reflected back into the initial medium. In certain conditions, however, a wave can be reflected inside a medium without splitting, retaining all of its intensity.

Total internal reflection requires a specific geometry – the wave must ricochet off the boundary at an angle greater than the critical angle θ_{crit}, defined by the refractive indices n_1 and n_2 of the two materials. The critical angle can be calculated from the equation:

$$\theta_{crit} = \arcsin \left(n_2 / n_1 \right)$$

Total internal reflection affects many different types of wave, and forms the principle behind optical fibres (shown opposite), used to transmit light signals in communications technology.

Diffraction

When a line of parallel waves passes through a narrow gap or is partially blocked by an obstacle, something unexpected happens – once the constricted waves have passed the obstruction, they start to spread out or bend. This is the most familiar form of a phenomenon known as diffraction – similar effects are created by light waves passing through materials with varying refractive indices, and by sound waves in analogous situations. Diffraction is best explained by Huygens' principle (see page 64).

Diffraction is a valuable tool for studying various aspects of waves, and light in particular. The diffraction patterns produced by light beams passing through two narrow slits and then interfering were a clinching piece of evidence for the wave nature of light. A series of narrowly spaced slits, meanwhile (a diffraction grating – see page 100), disperses light according to its wavelength, and is frequently used in optical devices in preference to a traditional prism.

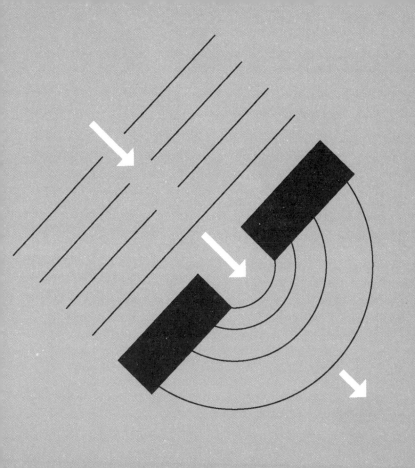

Resolution

The ability to distinguish fine detail and separate narrowly spaced objects is known as resolution. In technical usage it indicates the precision with which a variable in any kind of measurement can be determined. In astronomy and microscopy, optical resolution governs our ability to see tiny structures or separate binary stars, but is perhaps most familiar from pixels on a computer monitor.

The eye's ability to resolve (or be fooled by) these individually broken-down image elements depends on a general optical concept known as angular resolution – the narrowest angular separation at which two point sources of light can be seen as separate objects. For example, the average angular resolution of a human eye is around 1 minute of arc ($\frac{1}{60}$th of a degree). In principle, this depends on the interaction of diffracted waves as shown in the diagram opposite. The resolution of telescopes and microscopes depends on their specific design, but can be calculated using fairly simple equations.

1 Diffraction pattern created by light from point source 1
2 Diffraction pattern from point source 2
3 Limit of resolution where central peaks can be distinguished
4 Combined light from objects at limit of resolution

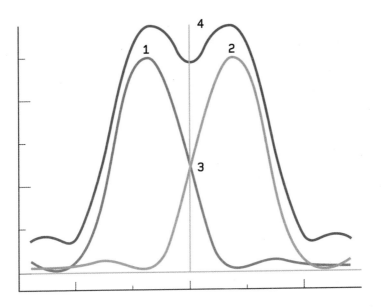

Scattering
and absorption

When waves interact with moving particles in a medium, they are affected by a variety of processes, collectively known as scattering, that tend to spread them out. Related absorption processes, meanwhile, diminish their energy.

Scattering of light can be either elastic or inelastic, depending on whether individual photons retain or change their energy. The most familiar form of elastic scattering, known as Rayleigh scattering, occurs when light interacts with particles that are significantly smaller than its own wavelength: the shorter the wavelength, the stronger the effect. Rayleigh scattering is responsible for the blue colour of the daylight sky and the yellow appearance of the Sun (turning red as the effect becomes more pronounced near sunset). A common form of inelastic scattering, known as Compton scattering, involves a transfer of energy from a photons (individual packets of electromagnetic waves – see page 142) to charged particles such as electrons (see page 162).

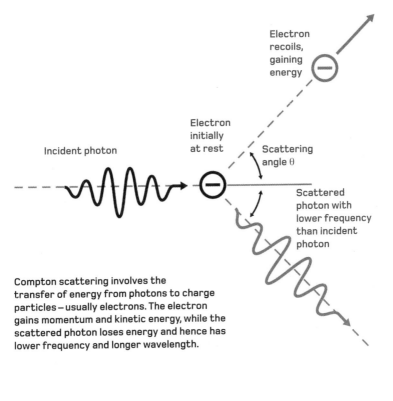

Electron recoils, gaining energy

Electron initially at rest

Incident photon

Scattering angle θ

Scattered photon with lower frequency than incident photon

Compton scattering involves the transfer of energy from photons to charge particles – usually electrons. The electron gains momentum and kinetic energy, while the scattered photon loses energy and hence has lower frequency and longer wavelength.

Polarization

Certain types of transverse waves – most notably light and other forms of electromagnetism – are capable of oscillating in a variety of different orientations. In other words, two waves emanating from the same source will not necessarily oscillate in the same plane. A wave's polarization is a measure of the angle of its oscillation – in the case of an electromagnetic wave it is the orientation of the wave's electric field component. However, the phenomenon is complex – a wave's polarization can be changed in an electromagnetic field, or even rotate as it travels through space.

Polarization is a useful property because it can reveal information about an original light source, or the structure of materials that light is reflected from or passing through. Polaroid filters (in which parallel crystals form a 'grille' that only allows through the component of light oriented in one direction) allow light with different polarizations to be selected and studied.

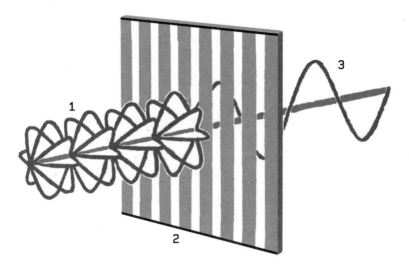

1 Unpolarized light naturally oscillates in multiple planes
2 A Polaroid filter acts as a narrow grille through which only
  correctly aligned components of light waves can pass
3 Emerging polarized light oscillates in just one plane

Lenses and prisms

A lens is a specially shaped piece of glass or another transparent medium that is used to divert the path of light rays, often in order to bend them to a focus. A lens typically consists of a piece of glass with two surfaces, one or both of which are ground and polished into a curved, symmetrical form around a central axis. Light rays passing through different parts of the lens strike it at different angles, and pass through varying depths of glass. Refraction in different parts of the lens causes the light rays to be bent to different degrees, and can be used to create various optical instruments (see pages 96 and 98).

A prism takes advantage of a phenomenon that other optical devices seek to avoid – the dispersion of light according to wavelength as it passes from one medium to another. It is a wedge-shaped piece of glass that seeks to maximize this dispersion, so that a single incident beam of light can be split up into a spectrum for study.

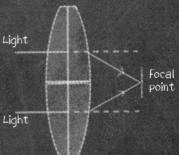

Concave
(divergent) lens

Light

Focal
point

Light

Convex
(convergent) lens

Light

Focal
point

Light

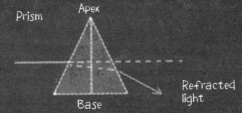

Prism

Apex

Light

Base

Refracted
light

Telescope

A telescope is an optical device for capturing light from distant objects and creating a magnified, more intense image. It takes advantage of the fact that light rays from distant objects are effectively parallel to each other, so that a single optical element (either a lens assembly in a refracting telescope, or a mirror with a curved surface in a reflecting telescope) can redirect all the light rays from a selected direction to a single point or focus.

As the light rays pass this point and begin to diverge once more, they can be intercepted by a smaller lens (or assembly of lenses) known as an eyepiece. The eyepiece typically refracts light rays onto a less sharply divergent path for observation by a human eye or capture by a camera or other detector instrument. The resulting image is both magnified (since the eyepiece effectively acts as a magnifying glass) and more intense thanks to the fact that the telescope has a much larger light-collecting area than the human eye.

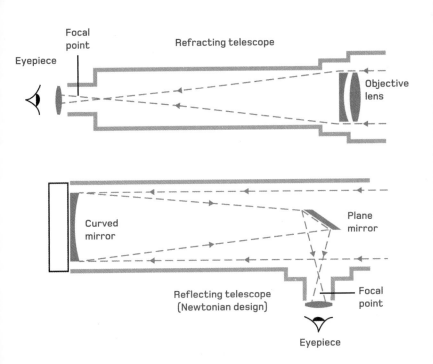

Refracting telescope

Focal point

Eyepiece

Objective lens

Curved mirror

Plane mirror

Focal point

Reflecting telescope
(Newtonian design)

Eyepiece

Microscope

In contrast to a telescope, a microscope is a device for producing magnified images of objects that are nearby and very small. The simplest form consists of just a single convex lens (a magnifying glass or loupe), which bends diverging light rays from the subject under study onto angles that are closer to parallel. The light therefore appears to diverge from a 'virtual image' with a larger angular diameter and apparent size than the original subject. More complex compound microscopes use multiple lenses – a powerful objective lens bends light to a focus inside the tube, forming a 'real image', and, as light diverges from this focus point, an eyepiece consisting of one or more lenses bends it back onto shallower and more parallel paths, once again forming an enlarged virtual image. Because microscopes are magnifying the light from relatively tiny objects, they require strong additional illumination (either from above the subject or, if it has been prepared as a narrow slice on a glass slide, from behind) in order to create a bright, well-contrasted image.

A compound microscope uses two convex lenses to form an enlarged virtual image. The objective is a 'strong' lens with a short focal length, while the eyepiece has a longer focal length and creates an image that appears to come from an optical 'sweet spot' 25 cm (10 in) from the eye.

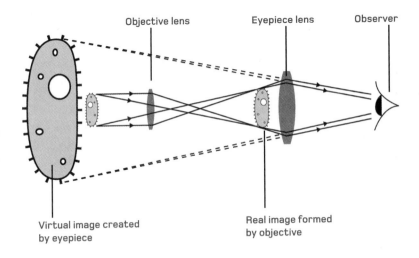

Objective lens

Eyepiece lens

Observer

Virtual image created by eyepiece

Real image formed by objective

Diffraction gratings

A diffraction grating is a device for splitting or dispersing light in different directions according to its wavelength. A grating typically consists of a finely spaced pattern of parallel lines splitting a reflecting or transmitting surface into separate linear areas. Diffraction and interference between light striking different parts of the grating disperses the light at different angles depending on its wavelenghth – the more tightly packed the lines, the greater the dispersion. Gratings are widely used in applications that require the creation of spectra or the separation of specific wavelengths, since they absorb less light than a glass prism, and can achieve greater dispersion angles. The principle was discovered by the 17th-century Scottish scientist James Gregory, based on studies of iridescent bird feathers. Gratings may be either reflective or transmissive – reflective designs may consist of a series of embossed lines on a mirror surface, while transmissive ones may use a pattern of opaque lines printed on transparent glass, or even a physical grille of finely spaced wires.

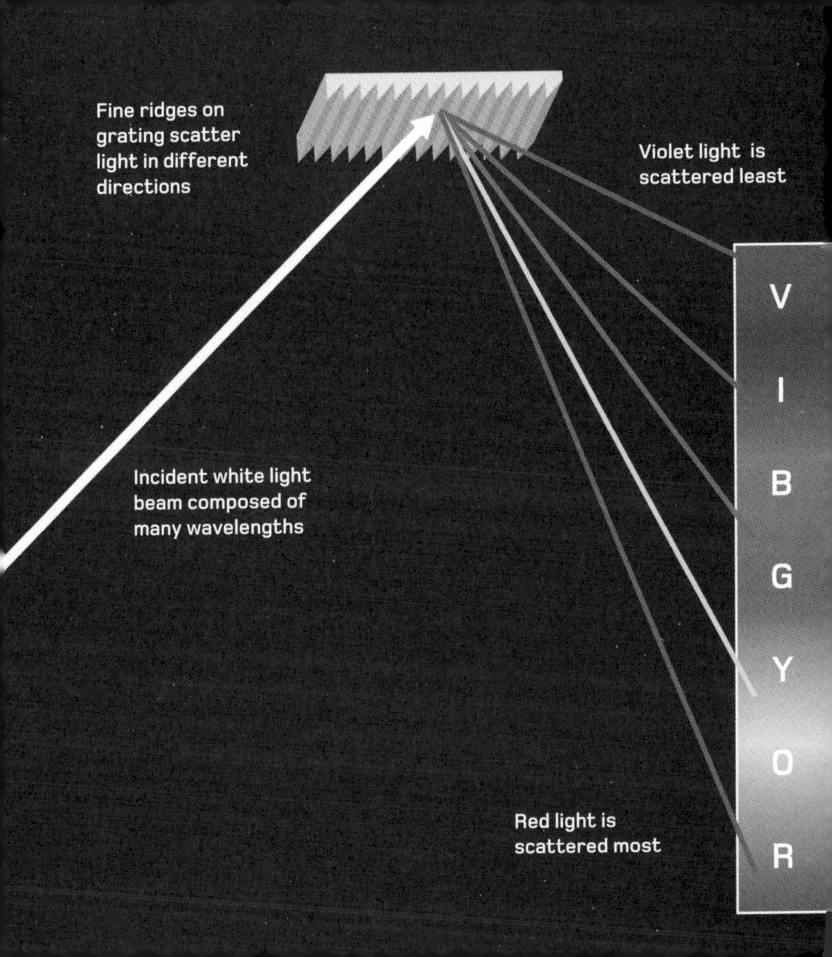

Fine ridges on grating scatter light in different directions

Violet light is scattered least

Incident white light beam composed of many wavelengths

V
I
B
G
Y
O
R

Red light is scattered most

Interferometry

The field of interferometry encompasses a number of different techniques, but all involve superimposing waves and studying the interference patterns they create in order to gather information. Although in theory it's possible to do interferometry with waves of any kind, in practice it almost always involves light or other electromagnetic waves. In most cases, a single coherent light beam from an original source is split into a 'reference beam' and a 'sample beam', often by passing them through a semi-silvered mirror. The sample beam is then modified in some way, before the two beams are recombined and allowed to interfere with one another.

Interferometers have a wide range of designs and functions – they can be used for precise distance measurement, as gyroscopes or rotation sensors, for testing of optical elements, for analysis of materials, and even as filters to block out unwanted radiations. They can even be used for investigating fundamental questions about the Universe (see page 384).

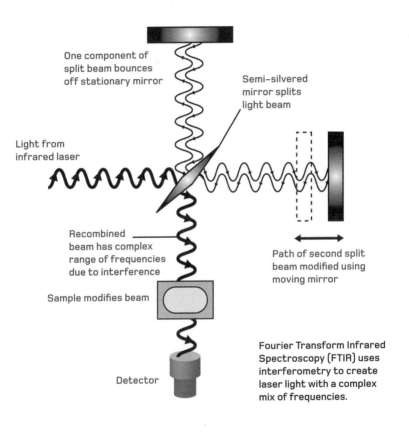

One component of split beam bounces off stationary mirror

Semi-silvered mirror splits light beam

Light from infrared laser

Recombined beam has complex range of frequencies due to interference

Sample modifies beam

Path of second split beam modified using moving mirror

Detector

Fourier Transform Infrared Spectroscopy (FTIR) uses interferometry to create laser light with a complex mix of frequencies.

Thermodynamics

Countless physical processes involve the transfer of energy between materials and locations in the form of heat – thermodynamics is the branch of physics that studies the properties of heat and the ways in which it can be transferred. While at first glance this might seem to be a highly specialized field with limited applications, it turns out to have important applications in almost every branch of physics, on scales that range from predicting the behaviour of individual atoms to the ultimate fate of the Universe.

Thermodynamics developed in the 18th and 19th centuries from the study of steam engines. Although such machines had been powering the Industrial Revolution for the best part of a century, the physical principles behind their operation were poorly understood until French scientist Nicolas Sadi Carnot developed his model of more generic 'heat engines' in the 1820s. Classical thermodynamics limits its considerations to the large-scale or 'macroscopic' properties of materials rather than the behaviour of their individual atoms. However, as with the kinetic theory of gases (see page 48), the essential laws of thermodynamics can be derived statistically by modelling the average behaviour of large numbers of particles.

The concepts of temperature and heat lie at the heart of thermodynamics. Heat is a form of energy that can be transferred between materials, while temperature is an innate – but changeable – property of materials. In statistical mechanics, temperature is interpreted as a reflection of the vibrational or kinetic energy within their individual particles – as a subtance's temperature increases, so does the *average* kinetic energy of its atoms or molecules.

Depending on internal structure, different materials respond to heat in different ways, as demonstrated by the fact that the same amount of heat applied to similar amounts of two substances may increase their temperature by different amounts. A material's response to the application of heat is described by a quantity known as its heat capacity (see page 116), while the transitions between phases of matter require the addition or removal of additional energy known as latent heat (see page 118).

Temperature and its measurement

On a microscopic level, a material's temperature is simply a manifestation of the average kinetic energy of its individual particles – whether they are trapped in the crystalline lattice of an iron bar, or floating freely in a cloud of steam. Most systems for measuring temperature rely on identifying a macroscopic property of the material that changes alongside its internal kinetic energy, selecting two points between which that property changes in a linear or at least mathematically simple manner, and establishing a scale between them.

Traditional mercury thermometers, for example, rely on the fact that the liquid metal expands linearly when exposed to temperatures between the freezing and boiling points of water (as kinetic energy is transfered into its own atoms). Electronic thermometers rely on changes to the conducting or resistive properties of certain materials, while pyrometers and infrared thermometers directly use the radiation released by an object in order to measure its temperature.

In cold conditions, slow-moving air molecules transfer little energy to the mercury in a thermometer, so it does not expand.

In hot conditions, air molecules move faster, and kinetic energy transferred into the mercury causes it to expand.

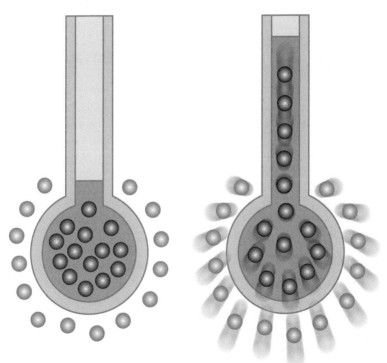

Absolute zero

Most thermometers and temperature scales are designed to measure the range of temperatures encountered in everyday life – for instance, the commonly used Celsius scale is based around the freezing and boiling points of water at standard pressure. But while for practical purposes there is no upper end to any temperature scale (since the kinetic energy of particles can be increased *ad infinitum*), there is definitely a bottom end. This is absolute zero – the temperature at which particles lose all their kinetic energy and stop moving.

Absolute zero is equivalent to −273.15°C (−459.67°F), and while it has so far proved impossible to achieve in a laboratory, scientists have come very close, discovering behaviours such as superconductivity and superfluidity in the process (see pages 250 and 320). British physicist Lord Kelvin was the first to suggest a temperature scale based on absolute zero – his Kelvin scale uses the same temperature intervals as the Celsius scale, so that 0°C is 273.15 kelvin (K), and 100°C is 373.15 K.

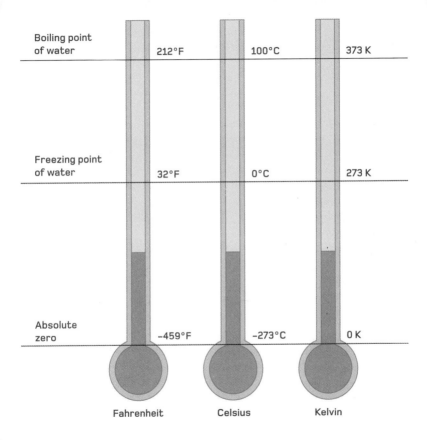

Heat transfer

The three principal ways in which heat can move through a material and be transferred from one location to another are usually described as convection, conduction and radiation.

Convection is the bulk movement of hot material through cooler surroundings, as seen in boiling water – all else being equal, hotter regions expand and become less dense, causing them to rise above denser ones. Conduction involves transfer of heat between particles on a microscopic scale. Collisions between fast-moving hotter particles and slower-moving cooler ones increase the kinetic energy and temperature of the cooler particles, but the particles themselves do not move in bulk through the heat-conducting medium.

Finally, radiation is the transfer of heat via electromagnetic rays. All materials are constantly emitting and absorbing radiation from their surroundings (see page 128), and this is the only form of heat transfer that can cross a vacuum.

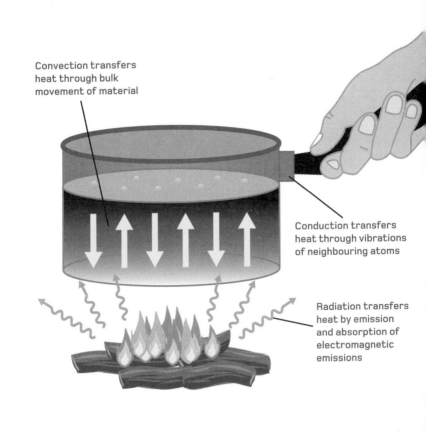

Convection transfers heat through bulk movement of material

Conduction transfers heat through vibrations of neighbouring atoms

Radiation transfers heat by emission and absorption of electromagnetic emissions

Enthalpy and entropy

Two related quantities are used to describe the distribution of energy in thermodynamic systems. Enthalpy (indicated by a letter H) is the total energy within a system, including that required to create it and to displace the environment around it. Although it cannot be measured directly, changes to enthalpy can be measured: when energy is added to a system, a process is said to be endothermic. When energy is removed, conversely, the process is exothermic.

Entropy (S), meanwhile, is a more complex property that defines the amount of thermal energy in a system that is unavailable to do mechanical work. On a microscopic scale, it is often taken as a measure of the amount of disorder within a system, and the degree to which energy is evenly distributed between particles with no thermodynamic imbalances that can be usefully 'put to work'. According to the second law of thermodynamics (see page 115), the entropy in a system inevitably increases unless external work is done to counter it.

Entropy causes an inevitable increase in the disorder of a system — if no external work is supplied to keep the system orderly, heat and energy will inevitably spread out to reduce thermodynamic imbalances.

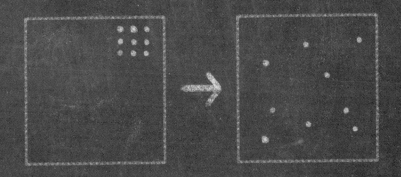

Laws of thermodynamics

Four fundamental laws of thermodynamics help define many of the field's important quantities including temperature, energy and entropy (see page 112).

The most essential law of all is, by historical accident, known as the 'zeroth law' (since it was only formalized after the other three). It states that if two objects are in thermal equilibrium with a third (in other words, no heat is flowing between them), then they must also be in thermal equilibrium with each other.

The first law states that both heat and work are forms of energy transfer, and that the internal energy of a closed system changes if heat is transferred in or out by work done to or by the system in a wider external context. More simply stated, doing work inevitably drains energy from even the most efficient of systems, and so perpetual motion machines (which allegedly can do work without expending energy) are forbidden by the laws of physics.

The entropy of an isolated system never decreases, and the second law states that any such system will inevitably evolve towards a state of thermodynamic equilibrium unless external work is applied to prevent it. Since thermodynamic equilibrium is the state at which entropy is at its greatest, the second law can also be elegantly stated by the phrase 'entropy increases'.

Finally, the third law of thermodynamics states that the entropy of a system approaches a constant value only as its temperature approaches absolute zero. The exact value of entropy depends on the quantum properties of the system involved, but for a pure crystalline substance, it too becomes zero at absolute zero.

The laws of thermodynamics not only help to define the terms involved in talking about thermodynamics, but also have deep cosmological and philosophical implications. The simple statement of the second law provides physics with an irreversible 'arrow of time', while the second and third laws together doom the entire Universe, as the largest isolated system of all, to a long, slow cooling and eventual 'heat death' – a state in which there are no temperature differences – unless, of course, other forces intervene in the meantime.

Heat capacity

A body's heat capacity is a measurement of the heat required to raise its temperature by a fixed amount. It is dependent on the material from which an object is made, and similar properties can be applied to materials themselves. Molar heat capacity is the heat capacity of 1 mole of a material (see page 50), while specific heat capacity is the heat capacity per unit mass of material, measured in units such as joules per kelvin per kilogram. Materials with low specific heat capacities tend to conduct heat rapidly, while those with higher capacities absorb it and act as insulators.

The heat capacity of a given substance is determined by how effectively heat energy is transferred into kinetic or vibrational energy of its individual atoms. Depending on its structure, there are various other ways in which energy can be absorbed *without* increasing kinetic energy – for example, as potential energy associated within the configuration of interatomic bonds or subatomic particles.

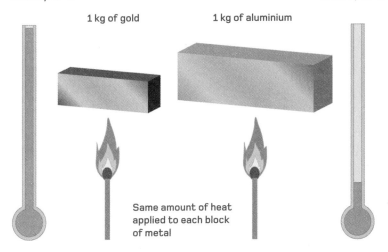

Temperature rises by 70°C

Temperature rises by 10°C

1 kg of gold

1 kg of aluminium

Same amount of heat applied to each block of metal

Due to their differing atomic properties, the specific heat capacity of aluminium is about seven times greater than that of gold (0.91 J/kg/°C compared to 0.13 J/kg/°C). As a result when the same mass of the two materials is subjected to the same amount of heat, gold heats up seven times more rapidly.

Latent heat

In most situations, adding heat to an object or system causes an increase in temperature, but this is not always the case. Sometimes, the supplied energy instead undertakes some other form of thermodynamic 'work' while temperature remains constant. The energy absorbed during such a process (and released when it is reversed) is known as its latent heat, or its enthalpy (see page 112), and the reconfiguration of matter involved is technically a 'phase transition'. The most familiar phase transitions are changes between states of matter such as melting or boiling (see page 156). Breaking and rearranging interatomic or intermolecular bonds during these processes requires an input of energy, and so they are described as endothermic. Condensation and freezing, on the other hand, release heat energy and are exothermic. The quantities involved in these two processes are known as the latent heat (or enthalpy) of fusion and vaporization respectively (typically measured in joules per kilogram), with the latter almost always the greater of the two.

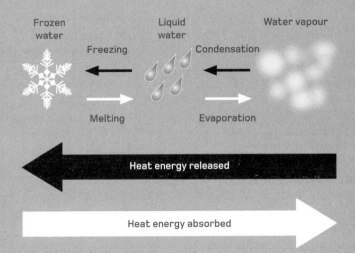

Heat engines

In the language of thermodynamics, any system that uses heat energy to perform mechanical work is known as a heat engine. In practice, heat engines function by raising the temperature of a 'working substance' and then using that substance to do work while excess heat is drained back into a colder 'sink', completing an overall 'heat cycle'.

In theory, heat engines can use any working substance, but in practice, gases or liquids (working fluids) are usually preferred. Steam engines are perhaps the classic example of this kind of device, doing work through the expansion and compression of the working fluid (water) as it evaporates during one stage of the cycle, and later condenses back into liquid. However, a physical phase change is not always necessary – internal combustion engines, for example, rely only on the expansion and contraction of gas. A refrigerator is an example of a 'heat pump' – an inverted heat engine in which mechanical work is used to transfer heat through a system and cool one part of it.

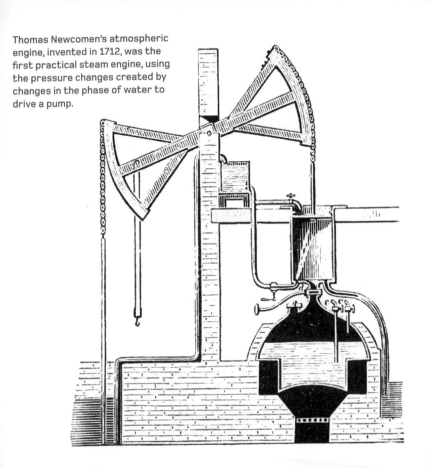

Thomas Newcomen's atmospheric engine, invented in 1712, was the first practical steam engine, using the pressure changes created by changes in the phase of water to drive a pump.

Carnot cycle

In 1823, French scientist Nicolas Léonard Sadi Carnot outlined the most efficient possible heat cycle for use in either a heat engine (converting thermal energy into work) or a heat pump (converting work into a heat differential). It involves a working gas sealed in a cylinder with a movable piston at one end, with heat supplied from a hot 'reservoir' and lost to a cool 'sink'. When operating as a heat engine, the cycle runs as follows:

• The gas expands 'isothermally' at a constant high temperature T_H, doing work on the piston as it does so. In this stage, its expansion is powered by absorption of heat energy and entropy from the reservoir.

• Expansion continues, but the gas now loses internal energy and cools. However, no energy is entirely lost from the system. Eventually it reaches the 'cold temperature' T_C.

• The piston now does work on the gas, so that it undergoes isothermal compression at T_C. In this stage, heat energy and entropy are lost to the sink.

• Finally, the compression continues, but now the gas increases its internal energy and its temperature returns to T_H.

The Carnot cycle is largely theoretical – it depicts an idealized system in which the cylinder and piston neither absorb nor emit heat themselves, and the reservoir and sink also maintain constant properties. No real heat engine can achieve such efficiency, but the Carnot cycle still provides a useful way of understanding the changes involved in various stages of any real-world heat cycle.

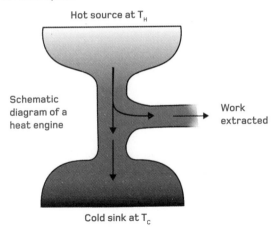

Hot source at T_H

Schematic diagram of a heat engine

Work extracted

Cold sink at T_C

Electromagnetism

Electromagnetic waves are one of the key ways in which we can learn about the Universe – they are emitted and absorbed by all types of matter that are familiar from our everyday experience, and transmit energy between objects even across the vacuum of space. The most obvious form of electromagnetism is visible light. This is a very small part of the wider electromagnetic spectrum, which our eyes have evolved to respond to – there's nothing inherently special about it, but it was studies of the properties of light that led to our broader understanding of electromagnetism and the discovery of invisible radiations such as infrared, ultraviolet and X-rays.

The precise nature of electromagnetism, however, has always been problematic. In most situations and experiments it displays the properties of a wave – for example it is subject to diffraction, refraction and interference. But it also travels through the vacuum of space without a medium to carry it, and in certain circumstances it can behave as if it is made up of individual particles or photons (see page 142). The most intuitive way of reconciling these differences is to imagine light and other forms of electromagnetism as self-contained bundles or 'packets' of energy that encapsulate wave-like properties within them – but the reality is even stranger (see page 288).

In most situations, however, we can treat electromagnetic waves as self-sustaining fields that move across space as transverse waves (see page 61), with their electrical and magnetic components perpendicular to each other, so that disturbance in the electric field reinforces the magnetic field, and vice versa. In a vacuum, these waves move at a speed of 299,792 kilometres per second (186,282 miles per second) – a natural constant called c that, remarkably, has proved to be the ultimate speed limit of the Universe (see page 366). Because the speed of electromagnetic waves is constant, it's easy to see that their wavelength and frequency are intrinsically linked, and inversely proportional to one another – the longer the wavelength, the lower the frequency and vice versa. The amount of energy carried by a wave depends crucially on its frequency, with higher-frequency waves, such as X-rays, carrying more energy than visible light, and lower-frequency waves, such as infrared and radio waves, carrying less.

Visible light

The light that is visible to our eyes all lies within a very narrow range of wavelengths, between 390 and 740 nanometres (nm). Its only special significance is that this is one of the few wavebands for which Earth's atmosphere is transparent, and therefore most life on Earth has evolved to take advantage of it through a variety of organs, most obviously the eyes.

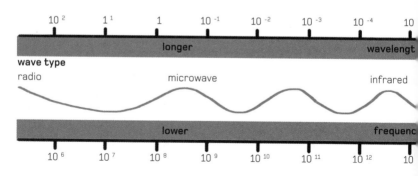

Isaac Newton proved in 1666 that the apparently pure white light that reaches Earth from the Sun is in fact a blend of different colours and wavelengths that can be split through a prism (see page 94). Processes such as diffraction and refraction affect different colours of light by different amounts on account of their different wavelengths. Visible light is only generated by relatively high-energy processes, and emitted by objects heated to temperatures of thousands of degrees. Nearly all natural light on Earth ultimately comes from the Sun.

The electromagnetic spectrum stretches from low-frequency radio waves to high-frequency gamma rays.

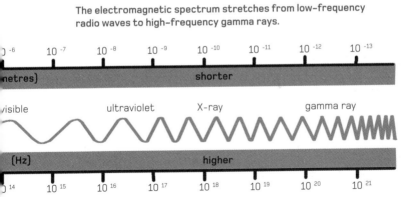

Infrared radiation

Electromagnetic radiation with wavelengths a little longer than the reddest visible light is known as infrared. Identified by German-born British astronomer William Herschel in 1800, it was the first invisible radiation to be discovered. During an experiment to measure the temperature associated with different wavelengths of sunlight, Herschel discovered that heat is still produced beyond the reddest visible light.

Infrared is defined as radiation with wavelengths between 740 nanometres and 1 millimetre. It is emitted by less energetic processes than visible light, and can be thought of as 'heat radiation'. Nearly all objects emit infrared, emitting frequencies related to their temperature (hot objects such as the Sun emit mostly high-energy near-infrared, while cooler objects such as the human body emit mid-infrared, and the coldest of all radiate only in the far infrared. These radiations have a huge range of applications, from thermal imaging devices, through wireless remote controls, to astronomical detectors.

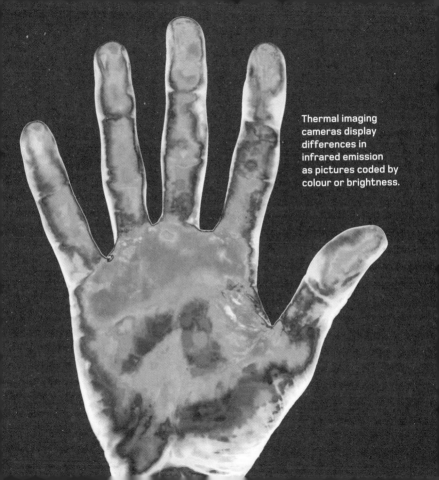

Thermal imaging cameras display differences in infrared emission as pictures coded by colour or brightness.

Radio waves

The existence of radio waves was predicted by Scottish physicist James Clerk Maxwell in 1867, based on his newly minted electromagnetic wave model for light and other radiations. Twenty years later, German scientist Heinrich Hertz succeeded in producing, transmitting and receiving them. Radio is the longest-wavelength, least energetic radiation, with wavelengths ranging from 1 millimetre at the 'short wave' end of the radio spectrum, up to hundreds of metres or even kilometres at the 'long wave' end. Frequencies, meanwhile, range from 300 gigahertz (300 billion wave cycles per second) at one extreme, to 3 kilohertz at the other.

Radio waves can be sent using an antenna in which an oscillating electric current generates an electromagnetic field (see page 228), and received with a similar antenna in which interaction with passing radio waves induces an oscillating current (see page 226). Using modulated radio waves (see page 68), this makes it possible to send information over long distances.

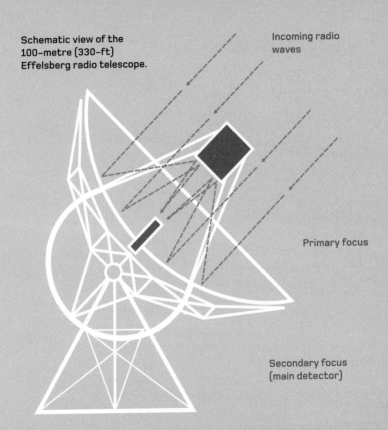

Schematic view of the 100-metre (330-ft) Effelsberg radio telescope.

Incoming radio waves

Primary focus

Secondary focus (main detector)

Microwaves

With wavelengths between 1 millimetre and 1 metre long, corresponding to frequencies between 300 gigahertz and 300 megahertz, microwaves are the shortest-wavelength, highest-energy form of radio waves. Their wavelength range makes them ideal for a number of applications, since they carry considerable energy and can be constrained to narrow beams. Applications include 'point-to-point' radio communications, radar and, of course, the microwave oven. Generating microwaves (especially those of the highest energies) requires the use of specialized devices such as magnetrons, in which fast-moving electrons are manipulated onto spiral paths by a magnetic field, in order to produce a high-frequency wave.

Microwaves are also widespread in nature, and are produced by a variety of astronomical processes (although these are mostly absorbed in Earth's atmosphere). Perhaps the most significant microwave signals are those generated by the decaying afterglow of the Big Bang (see page 400).

Schematic plan of a cavity resonator commonly used to generate microwaves.

Stable magnetic field

Electrons from hot filament spiral out to resonator

Currents around cavities generate microwaves

Ultraviolet radiation

Electromagnetic radiation with wavelengths a little shorter than visible light is known as ultraviolet or UV. Its existence was discovered in 1801 (just one year after the discovery of infrared – see page 128) when German physicist Johann Wilhelm Ritter found that light-sensitive silver chloride reacted more vigorously to rays from beyond the violet end of the spectrum than it did to visible light itself. Indeed, UV's abilities to trigger chemical reactions and cause substances to fluoresce in visible light are still at the heart of scientific and commercial applications, from forensics to hygiene and materials science.

Today, ultraviolet is defined as radiation with wavelengths from 10 to 400 nanometres. It forms a substantial part of the Sun's energy output, and can have harmful effects on human cells, by causing sunburn and triggering other chemical reactions. Fortunately, it is largely blocked by ozone in Earth's atmosphere (as illustrated opposite), but even so about 3 per cent of solar energy reaching Earth's surface still arrives in ultraviolet form.

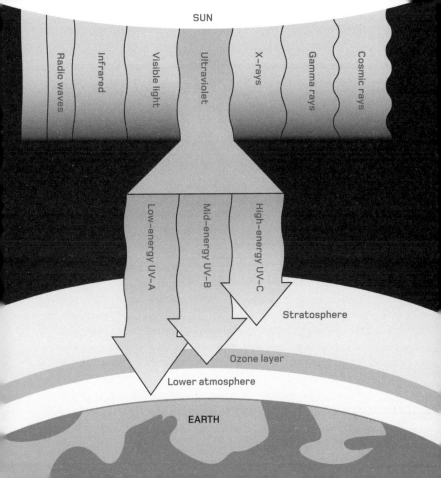

SUN

Radio waves

Infrared

Visible light

Ultraviolet

X-rays

Gamma rays

Cosmic rays

Low-energy UV-A

Mid-energy UV-B

High-energy UV-C

Stratosphere

Ozone layer

Lower atmosphere

EARTH

X-rays

These high-energy, penetrating rays were first recognized as a distinct type of radiation by German physicist Wilhelm Röntgen in 1895. Famously, while conducting experiments using Crookes tubes (a form of discharge tube similar to the cathode ray tube – see page 254), Röntgen discovered that some invisible form of radiation from his apparatus was 'fogging' his photographic plates, even when they were shielded by layers of black cardboard. Thus the science of X-radiography – using X-rays to look through permeable materials and image more opaque ones – was born.

With wavelenghs between 0.01 and 10 nanometres, X-rays pass through many materials unhindered, including conventional lenses and mirrors. Focusing and manipulating them usually requires special techniques, and their ability to damage human cells calls for limited exposure and heavy shielding. Fortunately, although many astronomical objects generate X-rays, Earth's atmosphere provides an effective absorbing shield.

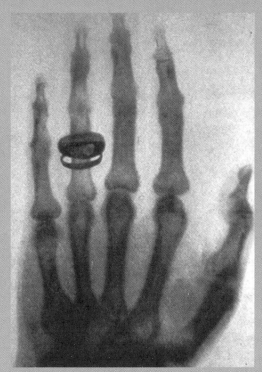

An early X-ray image made by Wilhelm Röntgen of Swiss anatomist Albert von Kölliker's hand.

Gamma rays

The shortest-wavelength and highest-energy radiations of all, gamma rays, were discovered around the turn of the 20th century by physicists investigating the processes of radioactive decay (see page 259). Gamma rays are often emitted during decay processes as individual radioisotopes shed excess energy. They carry huge amounts of energy and are powerfully ionizing – in other words, they can chemically alter any substances they strike. Conversely, however, their extremely short wavelengths (usually less than 0.01 nanometres, although precise definitions of the borderline between X-rays and gamma rays vary) allow them to pass through many materials without interaction.

Aside from radioactive sources, gamma rays can be produced by high-energy processes such as lightning, and in space by pulsars and black holes (see page 386). Despite being almost impossible to focus and manipulate, controlled gamma-ray sources have a number of uses in medicine and hygiene.

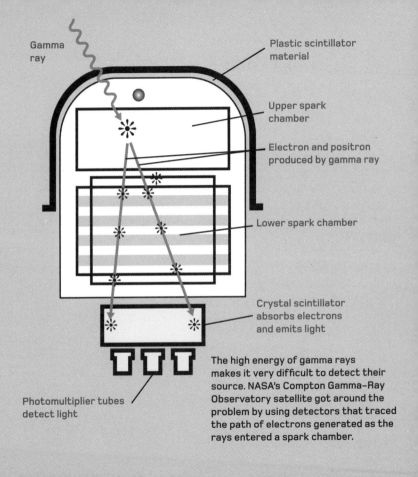

Gamma ray

Plastic scintillator material

Upper spark chamber

Electron and positron produced by gamma ray

Lower spark chamber

Crystal scintillator absorbs electrons and emits light

Photomultiplier tubes detect light

The high energy of gamma rays makes it very difficult to detect their source. NASA's Compton Gamma-Ray Observatory satellite got around the problem by using detectors that traced the path of electrons generated as the rays entered a spark chamber.

Spectroscopy

A huge range of natural and artificial processes generate electromagnetic energy of one sort or another, but few processes release equal amounts of energy evenly across all wavelengths or frequencies. Instead, the distribution of energy across the spectrum is often stronger or weaker at certain wavelengths. Spectroscopy is the study of such patterns.

A typical modern spectroscope uses a diffraction grating (see page 100) rather than a prism to project a spectrum from a narrow beam of light into either an eyepiece on a movable mount that allows precise measurement of the 'dispersion angle', or onto a detector such as a CCD camera (see page 256).

Spectra fall into three general forms: emission spectra, in which all the emitted radiation is constrained to a few tightly defined wavelengths; continuum spectra, in which radiation is emitted across a broad range of wavelengths (for example, black body radiation – see page 148); and absorption spectra, where one of the two forms of emission spectra is overlaid with well-defined dark 'absorption lines' due to intervening cooler gas. Continuum and emission spectra are typically emitted by burning or electrically excited materials, while absorption spectra are generated by lower-energy atoms and molecules (for example from cool intervening material) absorbing radiation of specific wavelengths out of a continuum or emission spectrum.

Both emission and absorption processes are linked to the atomic structures of the chemicals involved (see page 178), since they are caused by electrons within atoms absorbing or releasing energy to move between sharply defined energy levels. The resulting characteristic series of absorption or emission lines can be used as chemical fingerprints to detect the elemental and molecular composition of distant or inaccessible objects. The sharp definition of these series also gives physicists a convenient means of pinning down the specific wavelengths of light being emitted or absorbed, which can be very useful when an object's entire spectrum has been shifted in wavelength and energy by phenomena such as the Doppler effect (see page 72) or strong magnetic fields.

Photon

A photon is an elementary particle of light – the smallest divisible packet or 'quantum' of electromagnetic radiation itself. The idea that electromagnetic rays can be treated as a stream of particles when most of their behaviour is distinctly and uniquely wave-like may seem puzzling, but it is the only way to explain phenomena such as the photoelectric effect (see page 286). This duality now lies at the heart of quantum physics (see page 284), and the photon even does 'double-duty' as a carrier particle for electromagnetic force in theories of fundamental physics (see page 334).

From the point of view of electromagnetism, however, photons are best considered as small packets of radiation, travelling at the speed of light, c, and containing within them a set amount of energy defined by their wavelength or frequency. The total energy of a photon is given by the simple formula $E = h\nu$, where ν (the Greek letter 'nu') is the wave's frequency and h is Planck's constant, equivalent to 6.63×10^{-34} joule seconds.

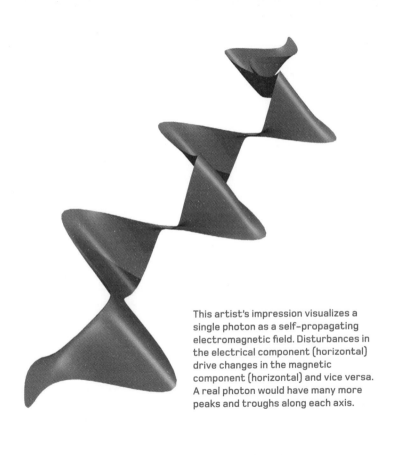

This artist's impression visualizes a single photon as a self-propagating electromagnetic field. Disturbances in the electrical component (horizontal) drive changes in the magnetic component (horizontal) and vice versa. A real photon would have many more peaks and troughs along each axis.

Laser

A laser (an acronym of Light Amplification by Stimulated Emission of Radiation) produces a beam of light in which all the individual waves are monochromatic (of a single wavelength) and coherent (in step with one another). This makes it an extremely powerful and intense light source that can deliver enormous amounts of energy onto a tightly focused spot.

Coherent light waves are achieved by stimulated emission, a process in which a suitable 'lasing medium' is pumped with energy (often using intense light or an electrical field) that would normally escape through spontaneous and disorderly fluorescence (see page 180). Before this can happen, however, the arrival of another photon jump-starts the emission process, forcing the lasing medium to emit another photon identical in every respect to the incoming one. The process repeats and cascades as light bounces back and forth in the medium, until it is eventually released in a tight beam. Uses of lasers range from surgery and manufacturing to precise measurement.

Light from a normal light source is a mix of wavelengths and frequencies, and spreads out in all directions.

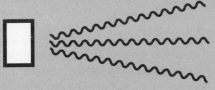

Light from a monochromatic source has a single wavelength, but its waves may not be in step with each other.

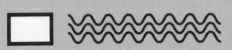

Light from a laser is monochromatic, coherent (with its waves in step) and aligned in a narrow beam.

Holography

Holography is an ingenious technique that uses lasers and the principles of interference and diffraction (see pages 66 and 86) to store information. It is probably most familiar through its ability to create uncanny three-dimensional images, but has important applications in fields ranging from security (since holograms are difficult to forge) to data storage (where they can be used to store and retrieve huge amounts of information at high speed).

A holographic image is effectively a recording of the 'light field' created around an object. Unlike a photograph, which stores the image itself, a hologram stores a pattern of varying density or intensity that can be used in one of several processes to reconstruct the original light field. Because the field records light scattered in various directions by the original object, it can also be viewed from various directions. The technique works only with monochromatic light sources such as lasers – with a mix of wavelengths, interference destroys the effect.

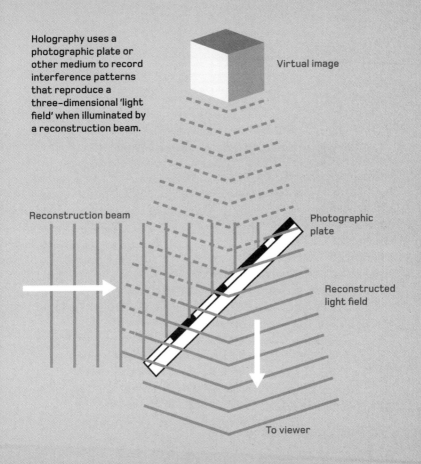

Holography uses a photographic plate or other medium to record interference patterns that reproduce a three-dimensional 'light field' when illuminated by a reconstruction beam.

Virtual image

Reconstruction beam

Photographic plate

Reconstructed light field

To viewer

Black body radiation

In physics, a 'black body' is an ideal object with a perfectly
light-absorbent, opaque and non-reflective surface, whose
emission of radiation is dependent on its surface temperature
and nothing else. Such an object might seem like a theoretical
abstraction, but in fact it is a good model for a wide variety of
phenomena ranging from stars to incandescent lamp filaments.
As a result, black body radiation was subject to intense study in
the mid-1800s. Scientists such as Scotland's Balfour Stewart
and Germany's Gustav Kirchhoff showed how black bodies
emit radiation across a range of wavelengths and energies, in
a distinct 'black body curve' (opposite) whose precise position
and shape depends only on the surface temperature according
to a relationship known as the Stephan-Boltzmann law. By
measuring energy from a black body at specific wavelengths,
it's possible to work out its surface temperature. However,
the cause of the black body curve distribution itself and, in
particular, its behaviour at ultraviolet wavelengths were
only resolved in the early 20th century (see page 288).

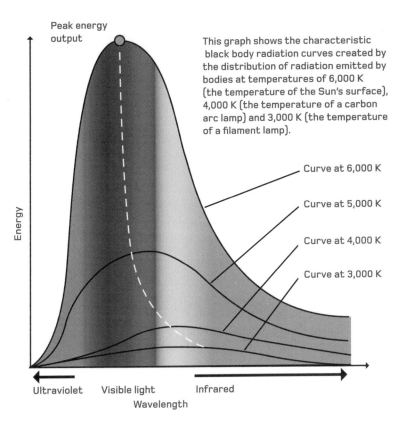

Peak energy output

This graph shows the characteristic black body radiation curves created by the distribution of radiation emitted by bodies at temperatures of 6,000 K (the temperature of the Sun's surface), 4,000 K (the temperature of a carbon arc lamp) and 3,000 K (the temperature of a filament lamp).

Curve at 6,000 K

Curve at 5,000 K

Curve at 4,000 K

Curve at 3,000 K

Energy

Ultraviolet

Visible light

Infrared

Wavelength

Cherenkov radiation

Although the speed of light in a vacuum, c, is the ultimate cosmic speed limit, light does move significantly more slowly in other transparent media. As a result, particles can sometimes achieve faster-than-light speeds in specific media. When this happens, they generate Cherenkov radiation, analogous to the sonic boom from supersonic aircraft.

The phenomeon requires a charged particle and a 'dielectric' medium (an insulating material whose structure can be rearranged or polarized by electric fields). The passage of the particle disrupts the electromagnetic field in the dielectric, causing its molecules to become briefly polarized before reverting to their normal state, emitting light in the process. When the particle is moving at normal speeds, destructive interference between the emitted radiation renders it invisible, but when the particle moves at high speed, light is emitted along a 'shock front' and interferes constructively to appear brighter and become visible, usually as a bluish glow.

Cherenkov radiation generated by fast-moving particles emitted around the core of a nuclear reactor.

Matter

The word 'matter' is commonly used to describe the substance from which normal physical objects are made. Matter is made up of particles, and displays mass and volume. But at its most basic, matter is a manifestation of energy. By its very nature, matter 'binds up' more energy than light and other forms of electromagnetic radiation, and so it accounts for the majority of energy in the visible universe. However, appearances can be deceptive: the normal matter of our everyday experience is vastly outweighed by mysterious, unseen 'dark matter' that only makes its presence felt through gravity (see page 402), while the energy in matter is now known to be dwarfed by equally mysterious 'dark energy' (see page 404).

In our everyday experience, however, matter is simply the tangible 'stuff of the universe'. It can display three classic phases or states – solid, liquid and gas – and many substances can transform between these three when their temperature is changed, either adding or removing energy from the system.

Solids, liquids and gases can display a wide range of properties and behaviours that depend on their composition at a more fundamental level; all large-scale matter is essentially composed of particles called atoms that are typically about 0.1 nanometres across – roughly one-thousandth the wavelength of visible light. Greek philosopher Democritus was the first to put forward an 'atomic theory' of nature around 400 BC (and indeed the word 'atom' comes from the Greek for 'indivisible') but it was only in the 18th century that evidence mounted to prove the existence of these tiny, submicroscopic particles. Today, we understand that atoms can be separated into scores of different elements. Each is unique in its overall properties, but most can be categorized into broad groups such as the metals and the noble gases. The complex ways in which atoms of different elements interact with each other is the basis for the entire science of chemistry, though we can only scratch the surface of those complexities here.

Meanwhile, 20th-century advances have also shown that atoms are far from the indivisible lumps that Democritus imagined – instead, they are composed of numerous subatomic particles: protons, neutrons and electrons, whose complex relationships help explain the properties the atoms themselves display.

States of matter

Matter is commonly found in one of three states – solid, liquid or gas. A 'fourth state', known as plasma, is primarily composed of atoms that have lost their outer electrons (see page 162). Solids maintain a fixed shape and volume regardless of any container they are in. Liquids maintain their volume but alter their shape to fit their container, while gases have no fixed shape and expand or compress to fill the available space.

Today, we understand that all three states are evidence of atomic or molecular-level behaviour. In a solid material, atoms and molecules are closely packed, and tightly bound together by strong interatomic and weaker intermolecular bonds, which often build up into a large-scale geometric or crystalline structure. When the material is heated, individual atoms can vibrate within the structure but not move through it.

In a liquid, atoms or molecules are still fairly tightly packed together, but the bonds between them are weaker, and can break and re-form with ease, so that the constituent particles can move past one another and rearrange their distribution. Finally, in a gas, the bonds linking individual atoms or molecules are broken and the forces between them are nearly negligible, allowing the particles in the substance to behave according to the gas laws (see page 46).

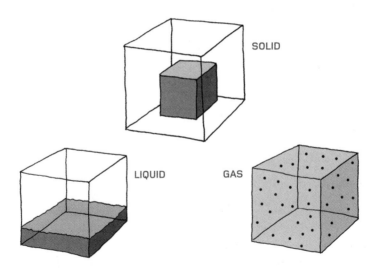

SOLID

LIQUID

GAS

Brownian motion

Most of the early evidence for atomic theory came from careful consideration of the gas laws (see page 46) and interpretation of experiments that showed how substances reacted with each other and formed products in relatively simple ratios. However, as early as 1827, botanist Robert Brown noted a strange effect – the seemingly random motions of pollen grains in water seen through his microscope.

Brown's experiments proved the phenomenon was not related to life, but the cause of 'Brownian motion' remained a mystery until 1905, when Albert Einstein realized that it could be explained in terms of invisible atoms ricocheting off the larger, visible particles. Einstein's elegant paper on the subject showed how Brownian motion could be used to calculate the size of the atoms or molecules in a liquid or gas. By extension, it also provided a method for calculating the number of atoms in a fixed volume of gas (see page 50), and therefore measuring the tiny masses of individual atoms themselves.

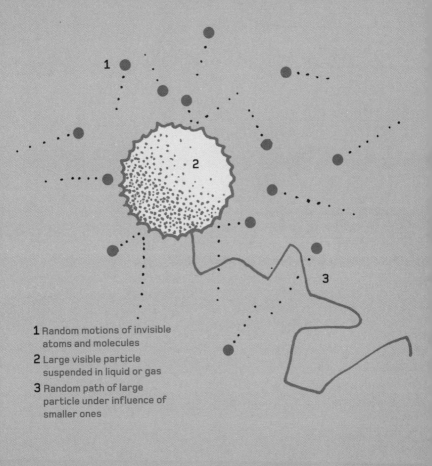

1 Random motions of invisible atoms and molecules

2 Large visible particle suspended in liquid or gas

3 Random path of large particle under influence of smaller ones

Chemical elements

A chemical element can be defined as a substance made from atoms of a single type, which displays its own unique set of physical and chemical properties. At present, some 118 of these basic units of matter are known – and 20 of these are the result of artificial experiments rather than natural discoveries. Each element has its own unique atomic number, an indicator of the internal structure of its atoms (see page 160). Elements also have an 'atomic mass' that generally increases alongside the atomic number (but see page 170).

In 1869, Russian scientist Dmitri Mendeleev laid out a list of the known elements and their properties and discovered some important patterns of distribution. This ultimately gave rise to the periodic table (see page 184), in which elements are listed in an order that reflects not only their increasing atomic number, but also the similarities of their chemical reactions. This, too, is now understood to reflect the internal atomic structure of the different elements.

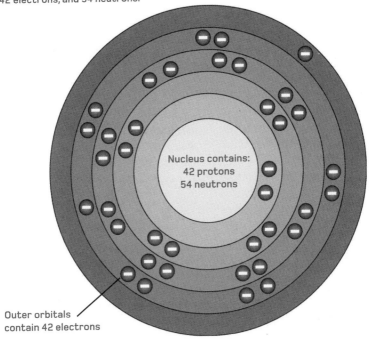

An atom of molybdenum has atomic number 42, and a mass number of 96, indicating that it contains 42 protons, 42 electrons, and 54 neutrons.

Nucleus contains:
42 protons
54 neutrons

Outer orbitals contain 42 electrons

Atomic structure

A typical atom has a diameter of around 0.1 nanometres, but its interior is largely empty space – nearly all of its mass is concentrated in a tiny core called the nucleus. This is composed of particles with positive electric charge called protons, and uncharged particles called neutrons, that have roughly equal masses, each equivalent to 1 atomic mass unit (amu). The 'mass number' of an element in the periodic table (see page 184) is therefore an indicator of the combined number of protons and neutrons in the nucleus. The atomic number, meanwhile, indicates the number of protons alone. In a neutral atom with no overall electric charge, the number of protons in the nucleus is balanced by an equal number of lightweight particles with the opposite charge, called electrons. They contribute very little to an atom's mass (each weighs just $\frac{1}{1,836}$ the mass of a proton) but occupy most of its volume, orbiting the nucleus at high speed in a series of nested shells known as orbitals. Interactions between electrons in an atom's outermost orbital are responsible for its chemical attributes.

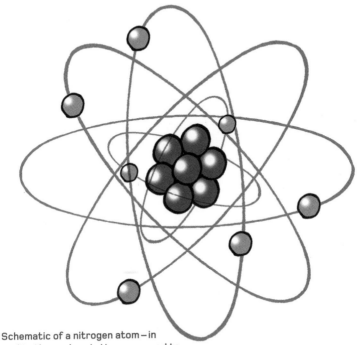

Schematic of a nitrogen atom – in
reality the nucleus is tiny compared to
the radius of the atom as a whole – typically
1/100,000th of its overall diameter.

Electron

The electron is one of three particles that make up matter atoms, and is the one most responsible for the chemistry of elements. In terms of classical physics, these tiny particles (a form of lepton – see page 330) orbit in shells around the atomic nucleus and occupy much of an atom's internal volume.

The negative charge of an electron is equal and opposite to the positive charge of a proton, so in a neutral atom, the number of electrons will precisely balance the number of protons (indicated by the element's atomic number). Electrons fill up specific niches known as orbitals extending from close to the nucleus outwards, with just two 'slots' for electrons in the innermost shell, eight in the second shell, 18 in the third, and so on according to a general formula where the number of electrons in shell n is $2n^2$. The mass of an electron is only $\frac{1}{1,836}$ that of a proton or neutron, so they contribute very little to an atom's overall mass, even in atoms with large numbers of electrons.

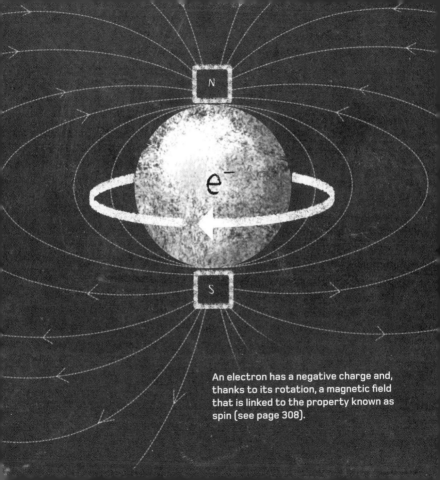

An electron has a negative charge and, thanks to its rotation, a magnetic field that is linked to the property known as spin (see page 308).

Proton

The proton is one of two relatively heavy particles that make up the nucleus of most atoms. Carrying a positive charge equal and opposite to that of the negatively charged electron, protons are composed of smaller particles called quarks (see page 328) – specifically two 'up quarks' and a single 'down quark'.

The number of protons in an atom's nucleus (its atomic number) defines the element that it belongs to. Although protons do not take part in chemical reactions themselves, the atomic number indicates the number and configuration of electrons in the outer shells of a neutral atom, which is critical to its reactivity. The simplest element, hydrogen, consists of a single proton orbited by a single electron – a naked proton, with no electron, is a hydrogen ion (H^+). A single proton weighs the same as 1,836 electrons – approximately 1.67×10^{-27} kilograms or 1 atomic mass unit (amu) – and has a diameter of about 1.7 femtometres (millionths of a billionth of a metre).

Hydrogen is the simplest atom, consisting of a single proton orbited by a single electron.

Neutron

The neutron was the last of the three principal subatomic particles to be discovered (as late as 1932, by English physicist James Chadwick) largely on account of its lack of an electrical charge. It has the same mass as a proton and is also usually confined within the central compact nucleus of an atom. However, in contrast to the proton, the neutron is composed of two 'down quarks' and a single 'up quark' (see page 328).

Neutrons contribute mass to the atomic nucleus, and are usually found in approximately the same numbers as protons. However, the number of neutrons can vary slightly between atoms of a single element without significantly affecting its chemical properties. Atoms with the same number of protons but different numbers of neutrons (and overall masses) are called isotopes. When the number of neutrons in a nucleus is significantly larger than the number of protons, however, the nucleus can become unstable and undergo radioactive decay (see page 260) in order to reach a more stable configuration.

An atomic nucleus consists of protons and neutrons in roughly equal numbers. Excess quantities of neutrons create heavier, unstable isotopes.

Antimatter

Despite its science fiction overtones, the concept of antimatter is actually rather simple. Antimatter particles are matter particles with the same mass, but opposite charge to their 'normal' counterparts. Hence, an antiproton is a particle with the mass of a proton but negative charge, while an antielectron, or positron, is an electron-like particle with positive charge. The possible existence of antimatter was demonstrated by British theoretician Paul Dirac in 1928, and positrons were identified by US physicist Carl Anderson in 1932, based on studies of 'cosmic rays' entering Earth's atmosphere.

Positrons are the most common type of antimatter, generated by radioactive decay and violent astronomical events. They often give away their presence from gamma radiation released as they come into contact with normal electrons and both particles 'annihilate' in a burst of energy. However, the overall question of why the Universe is dominated by matter rather than antimatter remains a major challenge for modern physics.

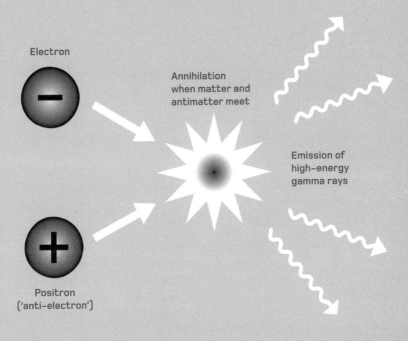

Electron

Annihilation
when matter and
antimatter meet

Positron
('anti-electron')

Emission of
high-energy
gamma rays

Isotopes

Isotopes of an element are atoms that share the same atomic number, but have slightly different masses thanks to different numbers of electrically uncharged neutrons (see page 166) in their atomic nucleus. Many elements naturally exist as a mix of different isotopes, although often one isotope is far more common than the others. In rare cases, the natural proportions of different isotopes are evenly balanced, resulting in an average atomic mass that may not be a whole number (for instance, the atomic mass of chlorine is 35.5).

Isotopes are significant because atomic nuclei with an excess of neutrons tend to be unstable, breaking down on short or long timescales through radioactive decay (see page 259) until they eventually reach a stable form. Different isotopes display a wide range of 'half-lives', and many of the shorter-lived forms are constantly regenerated in Earth's atmosphere and surface by interactions between atoms and high-energy particles called 'cosmic rays' from the Sun and deep space.

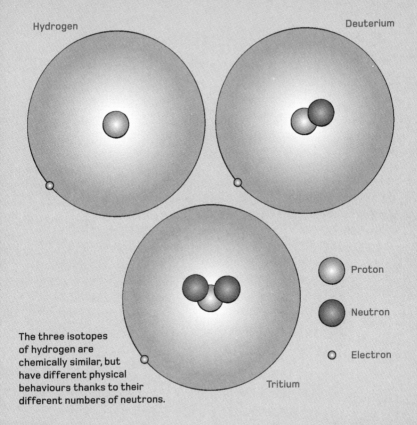

Hydrogen

Deuterium

Tritium

Proton

Neutron

Electron

The three isotopes of hydrogen are chemically similar, but have different physical behaviours thanks to their different numbers of neutrons.

Mass spectrometry

A mass spectrometer is an ingenious machine that allows scientists to analyse the precise isotopic composition of materials. It works by breaking down a substance into electrically charged ions and then firing them through a device that deflects them onto different paths according to their charge/mass ratio. The deflected ions then strike a detector plate, creating a 'spectrum' analogous to the one produced when light is split through a prism.

Samples can be broken into ions in a variety of ways, but one of the most common is electron ionization – bombardment with high-energy electrons from a heated electrical filament (see page 254). Similarly, the ions can be separated in different ways – using electric fields or magnets, or simply by measuring the speed at which different particles (subjected to the same acceleration) hit the detector. In all cases, however, the principle is the same: the intensity of particles in different parts of the spectrum reveals the isotope ratios of the original sample.

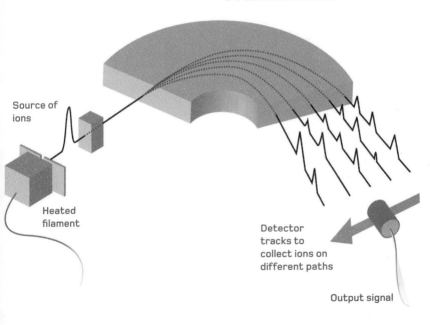

Magnetic or electric field deflects ions onto different paths. Particles with the highest charge/mass ratio are deflected the most.

Source of ions

Heated filament

Detector tracks to collect ions on different paths

Output signal

Ionization

By definition, an atom is electrically neutral. An atom-like particle that displays an electric charge is known as an ion, and the process of transforming an atom into an ion is known as ionization. Ions may be either positively or negatively charged, and their overall charge depends ultimately on the balance of positive protons in the nucleus and negative electrons in their outer orbital shells. Because the nucleus is tightly bound, its protons are rarely disturbed, so ions form by changes in the numbers of more easily influenced electrons.

A process that strips away electrons effectively removes negative charge, leaving an ion with a net positive charge known as a cation, while a process in which an atom gains an excess of electrons creates a negative ion known as an anion. During chemical reactions, both phenomena occur, although the ions often have only a brief independent existence (see page 192). Ionization can also be triggered by electromagnetic fields or by bombardment with high-energy radiation or particles.

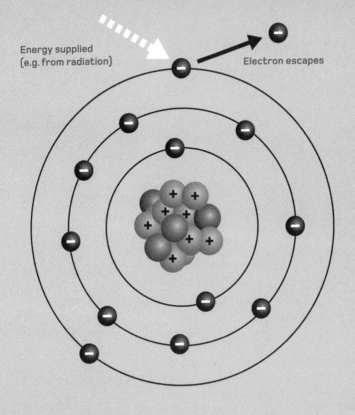

Energy supplied
(e.g. from radiation)

Electron escapes

Emission and absorption spectra

An emission spectrum shows the distribution of energy released when a particular substance is excited (injected with energy) to the point where it emits light – for instance, by burning in a flame or bombardment with energetic particles. When this light is split with a spectroscope (see page 140), most substances turn out to emit radiation in a series of emission lines with tightly defined wavelengths, energies and colours. Conversely, when incandescent light (a continuous range of colours – see page 148) is passed through a cooler, vaporized sample of the same material, light of the same specific wavelengths is absorbed, leaving a series of dark 'absorption lines'. Around 1859, German chemists Robert Bunsen and Gustav Kirchhoff realized that each element or compound produces a unique pattern of emission or absorption lines – a chemical fingerprint that allows it to be identified. Introduced in 1913, the Bohr model of the atom (see page 178), whose electron orbitals correspond to different energy levels, explains the unique behaviour of each element.

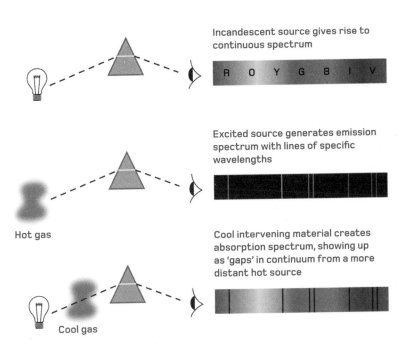

Incandescent source gives rise to continuous spectrum

R O Y G B I V

Excited source generates emission spectrum with lines of specific wavelengths

Hot gas

Cool intervening material creates absorption spectrum, showing up as 'gaps' in continuum from a more distant hot source

Cool gas

Bohr model of the atom

The early 20th century saw a series of important advances in our knowledge of atomic structure, culminating with the atomic model introduced by Danish physicist Niels Bohr in 1913. Bohr's chief breakthrough was to explain the origin of spectral lines (see page 176) created when an atom absorbs or emits radiation with very specific wavelengths and energies.

Bohr's model describes a positively charged atomic nucleus around which electrons move in shell-like regions, today known as orbitals or energy levels, which can each accommodate a fixed number of electrons. An electron's location gives it a unique energy, and electrons fill up shells from the inside outwards. If the outermost shell is incomplete, then an electron from the next shell down may absorb energy and make the jump into the outer shell, creating an absorption line. This configuration is unstable, however, and the electron rapidly returns to its lowest available energy state, releasing energy of an identical wavelength to create an emission line.

In the Bohr model of an atom, electrons orbit the atomic nucleus in sharply defined 'orbitals' – they can only move between these levels by absorbing or emitting energy.

1 Electron absorbs energy and moves to a higher orbital

2 Electron emits energy in order to fall to a lower orbital

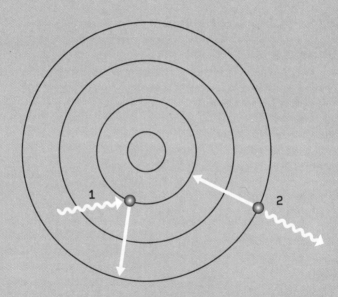

Fluorescence

Certain substances are capable of emitting light without being heated – instead their atoms become energized or excited by light or other radiations, then return to their lowest energy state through the release of light (in accordance with Bohr's model of the atom – see page 178).

This phenomenon, known as fluorescence, is most impressive when the excitation is powered by invisible (usually ultraviolet) light and only the emitted radiation is visible. This is the mechanism that illuminates the beautiful interstellar gas clouds known as emission nebulae, but it also lies behind more mundane phenomena such as fluorescent lamps. In these, electricity excites mercury vapour, forcing it to emit ultraviolet radiation that in turn excites a fluorescent material known as a phosphor. Unlike the broad spectrum of black body radiation (see page 148) emitted by incandescent sources, fluorescence typically emits only a limited range of wavelengths depending on the energy of the electron transitions involved.

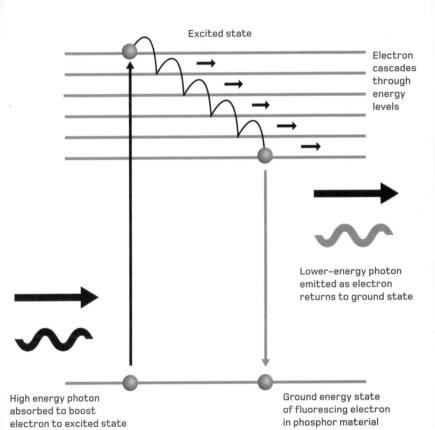

Excited state

Electron cascades through energy levels

Lower-energy photon emitted as electron returns to ground state

High energy photon absorbed to boost electron to excited state

Ground energy state of fluorescing electron in phosphor material

Atomic clock

When electrons in orbit around an atomic nucleus are excited by the injection of energy, they tend to return to their original, lower-energy state very rapidly, and the speed of return from a particular transition, if other conditions can be precisely controlled, is always identical.

This is the theoretical basis for atomic clocks, devices that provide the most accurate time measurement in the world today. Atomic clocks define time more accurately than even the rotation of the Earth itself – so precisely that, since 1967, the second has been officially defined as the duration of 9.192,631,770 cycles of radiation from a particular electron transition in atoms of caesium-133.

In practice, an atomic clock's time signal is produced by an electronic oscillator circuit that is driven at high frequency by magnetic resonance with gas atoms contained in a microwave cavity – a chamber in which electromagnetic fields are confined and amplified by resonance (see page 132).

Atoms are injected into the cavity in the form of a vapour, and a laser with a very precisely tuned freqency and wavelength is then fired at them. Each photon in the laser beam carries exactly the right amount of energy to trigger the desired transition in the electron energy state of the atoms, and as soon as an atom has returned to its lower energy state, it is boosted again by another laser photon. A resonant electro-magnetic field rapidly builds up in the resonator, driving the electronic timing circuit.

The transitions used in atomic clocks may release energy at either microwave, visible or ultraviolet wavelengths, and commonly involve atoms of hydrogen, caesium or rubidium. Although the principle behind these machines is well established, advances are still being made in their design and accuracy. For example, the most precise atomic clocks now cool their atoms to temperatures close to absolute zero (see page 108) in order to reduce unwanted variations in the transition frequency.

The periodic table

The periodic table originated as a simple means of ordering elements according to their chemical reactivity, but today it is also recognized as a means of describing the internal properties of atoms of particular elements. The table consists of 18 vertical columns of varying height, with elements increasing in atomic mass from left to right and top to bottom. Elements in these columns share similar chemical reactions.

Since chemical reactivity is linked to the number of electrons in an element's outer orbital shell (see page 160), it's no surprise to find that the arrangement of columns reflects the arrangement of these shells. For example, highly reactive elements with a single electron in their outer shells (including hydrogen, lithium, sodium and potassium) form a column at the extreme left of the table, while the unreactive 'noble gases' with full outer shells (including helium, neon and argon) lie in a column at the extreme right.

The periodic table of elements

1	2	3	4	5	6	7	8	9	10	11	12	13	14	15	16	17	18
1 H																	2 He
3 Li	4 Be											5 B	6 C	7 N	8 O	9 F	10 Ne
11 Na	12 Mg											13 Al	14 Si	15 P	16 S	17 Cl	18 Ar
19 K	20 Ca	21 Sc	22 Ti	23 V	24 Cr	25 Mn	26 Fe	27 Co	28 Ni	29 Cu	30 Zn	31 Ga	32 Ge	33 As	34 Se	35 Br	36 Kr
37 Rb	38 Sr	39 Y	40 Zr	41 Nb	42 Mo	43 Tc	44 Ru	45 Rh	46 Pd	47 Ag	48 Cd	49 In	50 Sn	51 Sb	52 Te	53 I	54 Xe
55 Cs	56 Ba		72 Hf	73 Ta	74 W	75 Re	76 Os	77 Ir	78 Pt	79 Au	80 Hg	81 Tl	82 Pb	83 Bi	84 Po	85 At	86 Rn
87 Fr	88 Ra		104 Rf	105 Db	106 Sg	107 Bh	108 Hs	109 Mt	110 Ds	111 Rg	112 Uub	113 Uut	114 Uuq	115 Uup	116 Uuh	117 Uus	118 Uuo

57 La	58 Ce	59 Pr	60 Nd	61 Pm	62 Sm	63 Eu	64 Gd	65 Tb	66 Dy	67 Ho	68 Er	69 Tm	70 Yb	71 Lu
89 Ac	90 Th	91 Pa	92 U	93 Np	94 Pu	95 Am	96 Cm	97 Bk	98 Cf	99 Es	100 Fm	101 Md	102 No	103 Lr

Ionic bonds

One of three common types of chemical atomic bond, an ionic bond, as its name suggests, involves the formation of electrically charged ions. They usually require the presence of a metal atom (an element whose relatively low 'ionization energy' allows electrons to be easily removed from its outermost orbital shell) and a non-metallic substance with an electron-accepting 'gap' in its outer orbital. During ionic bonding, an injection of energy strips the outer electrons from the metal atom, causing it to form a positively charged ion known as a cation. The electrons are then absorbed into the outer shells of one or more non-metallic atoms to complete or further stabilize their outer shells, creating a negatively charged ion known as an anion. Electrostatic forces (see page 204) attract the anions and cations in the substance to each other, forming a strong bond that, across a large volume of the solid substance, often displays a crystalline or lattice-like structure. The overall process of forming a bond must be exothermic (releasing rather than absorbing energy).

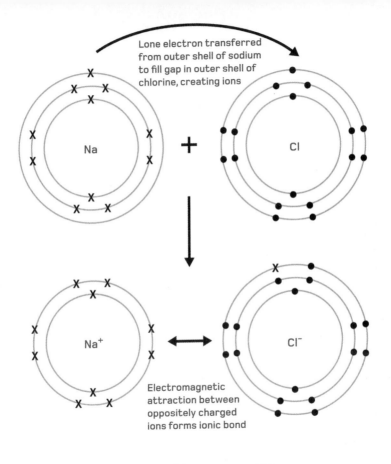

Lone electron transferred from outer shell of sodium to fill gap in outer shell of chlorine, creating ions

Na + Cl

Na$^+$ ⟷ Cl$^-$

Electromagnetic attraction between oppositely charged ions forms ionic bond

Covalent bonds

Alongside ionic bonds (see page 186), covalent bonds are one of the main ways in which atoms join together. They form in situations where two or more atoms need to gain additional electrons in their outer orbital shells in order to reach a stable configuration, and essentially involve the sharing of pairs of electrons between atoms – each atom donates one electron, and effectively gains one from from its partner in the bond.

A simple example of a covalently bonded molecule is methane (CH_4), in which a carbon atom with four electrons in its outer shell forms links with four separate hydrogen atoms, each of which has a single electron. By sharing the electrons across covalent bonds, the carbon atom achieves a stable arrangement of eight electrons in its outermost shell, and the hydrogen atoms achieve a similarly stable arrangement of two electrons in their outer shells. Precise descriptions of covalent bonds, explaining why they vary in strength and orientation, can only be achieved through quantum mechanics (see page 300).

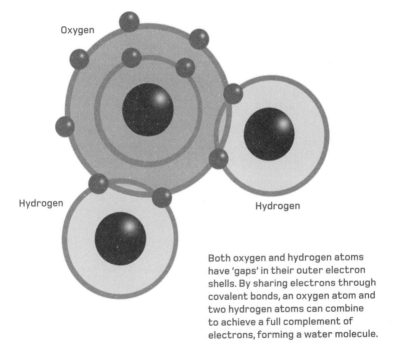

Oxygen

Hydrogen

Hydrogen

Both oxygen and hydrogen atoms have 'gaps' in their outer electron shells. By sharing electrons through covalent bonds, an oxygen atom and two hydrogen atoms can combine to achieve a full complement of electrons, forming a water molecule.

Metallic bonding

When metal atoms join together in bulk, a special type of bond regularly holds their structures together. Metallic bonding can be seen as a form of ionic bonding (see page 186), in which all the ions are positive, but they are held together by a 'sea' of delocalized electrons. The metal atoms surrender electrons from their outer orbital shells to become more stable, building up a regular lattice or crystalline structure, while the electrons 'float' between them, neutralizing the overall charge.

This unique bonding is responsible for many characteristic properties of metals. Mobile electrons flow easily across the lattice, making them good electrical conductors. This same mobility, combined with vibrations of atoms in the crystal lattice, also allows them to conduct heat well. The ability of the atoms to rearrange themselves under stress, meanwhile, accounts for the malleability and ductility of metals, while interactions between light rays and free electrons are responsible for their reflective, lustrous surfaces.

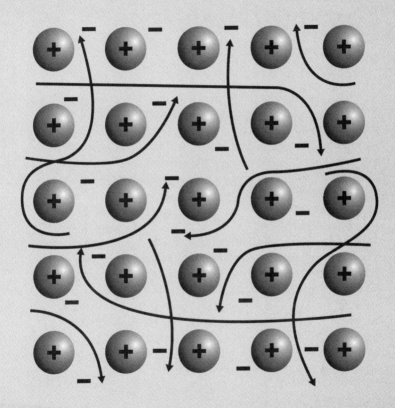

Chemical reactions

A reaction is the process by which one set of atoms and molecules, known as the reactants, is transformed into another set, known as the products. Chemical reactions typically involve the breaking and re-forming of atomic bonds and the redistribution of electrons between the reactants, but leave the atomic nuclei essentially unaltered.

As a rule, breaking chemical bonds absorbs energy, while creating new ones releases energy. As a result many reactions require the addition of energy at the start (to break down the reactants), but generate energy of their own as they progress (as the chemicals form new ionic or covalent bonds). If a particular reaction absorbs energy overall, it is described as endothermic, and if it releases energy on balance, it is exothermic. Reactions that start on their own, simply by introducing the reactants to one another (for example, the formation of sodium hydroxide when sodium is added to water) are said to be spontaneous or 'exergonic'.

Chemical reactions are often described using equations, such as:

$$2\,Na + 2H_2O \longrightarrow 2\,NaOH + H_2$$

By convention, the two sides of the equation are balanced so that the same number of atoms lies on each side, and the final products are indicated correctly (so in this example, the number of molecules on the left-hand side of the equation is doubled so that a balanced equation gives the molecular hydrogen (H_2) that is actually produced, rather than unstable atomic hydrogen.

A two-headed arrow in the middle of the equation indicates that a reaction is reversible – given the right conditions, the reaction will run in reverse, reclaiming the reactants from the products. In many cases, this is a purely theoretical possibility, but in many others, reversibility is an important factor: the reaction may run in either direction depending on the relative concentrations of reactants and products. This situation can result in an 'equilibrium' between reactants and products, although in theory any reaction that is allowed to run to completion reaches an equilibrium – even if that equilibrium involves all the reactants being converted into products.

Intermolecular bonds

A variety of forces are responsible for holding molecules together in solid and liquid states. Although generally weaker than the interatomic bonds *within* molecules, they play a vital role in shaping the bulk properties of matter. Frequently, these intermolecular bonds involve electrostatic forces of attraction (and repulsion) between concentrations of electric charge in neighbouring molecules. When charge is unevenly distributed in a molecule (often due to shared electron pairs formed by covalent bonding) it is said to be 'polar', and when a concentration of negative charge at one end of a molecule leads to a positive charge at the other, the result is a 'dipole'. The formation of dipoles provides several different possibilities for attraction between molecules. In broad terms, these may involve attraction between a dipole and an ion, or attraction between oppositely charged dipoles (so-called Van der Waals' forces). The influence of dipoles on neighbouring molecules can even polarize normally non-polar molecules.

Another important type of intermolecular bond is the hydrogen bond. This kind of bond forms when an electron-enriched atom attracts positively charged hydrogen nuclei from neighbouring molecules. It is particularly significant in molecules such as H_2O, and is largely responsible for water's melting and boiling points, which are unusually high when compared to similar molecules such as ammonia (NH_3).

Polarity of a water molecule

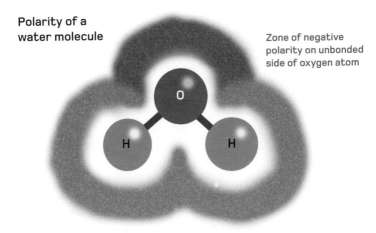

Zone of negative polarity on unbonded side of oxygen atom

Zones of positive polarity around hydrogen atoms

Solutions

In addition to the states of matter discussed on pages 152–3, many chemical compounds can form homogeneous (uniform) mixtures known as solutions. The process normally involves molecules of a solid, liquid or gas (the solute) becoming separated and distributed through a liquid known as a solvent. Intermolecular attractions to solvent molecules (see page 194) overcome the strength of the original bonds within the solute.

When ionic compounds dissolve, the interatomic bonds holding them together may be broken, allowing separate atoms or groups of atoms to form ions in the solution. A common example is table salt (sodium chloride), which breaks down to form positively charged sodium ions and negatively charged chlorine ions. The existence of these free-floating charged particles allows such solutions to conduct electricity – a phenomenon that is useful in a range of applications from separation or 'electrolysis' of chemicals to energy generation from electrochemical cells (see page 214).

A solution of common table salt consists of sodium and chloride ions surrounded by water molecules.

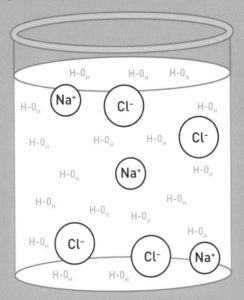

X-ray crystallography

For all the technological breakthroughs of the last few decades, the direct imaging of individual atoms and molecules still remains largely beyond our reach. However, for some materials at least, X-ray crystallography offers the next best thing – a detailed map of molecular structure.

The technique involves firing a beam of X-rays at or through a crystalline structure, and analysing the resulting diffraction pattern (similar to the one shown opposite). The crystal effectively acts as a diffraction grating for X-rays (see page 100). The technique was invented by German physicist Max von Laue in 1912, and developed by British father and son William Henry and William Lawrence Bragg, who worked out exactly how structure within a crystal influenced the angles of diffracted rays. In the past century, it has been used to interpret the molecular structure of a wide range of materials – perhaps most famously the double-helix shape of the DNA molecule – and has even been employed by robotic rovers on Mars.

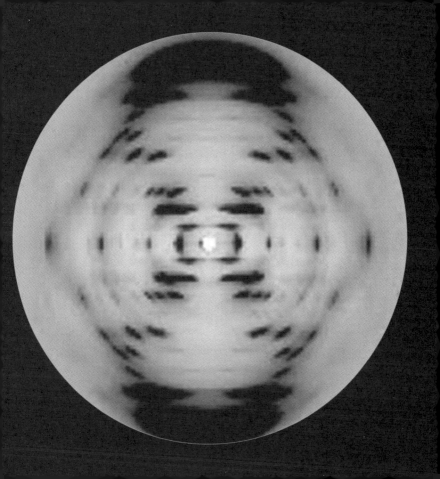

Atomic force microscopy

One surprisingly accurate form of microscopy is, so far, the only way in which we can image and manipulate matter on the scale of individual atoms and molecules. The idea behind the atomic force microscope (AFM) is surprisingly simple – an extremely sharp probe mounted on a cantilevered support is drawn across the surface of a material sample, and allowed to move freely up and down in response to forces of repulsion and attraction. By detecting these tiny movements (usually through minute deflections to the path of a laser beam reflecting off the top of the probe), a computer can reconstruct the atomic-scale features of the surface itself.

What's more, the AFM is not merely an observing tool – by applying a force to the probe it can be used to push atoms and molecules around. So far the technology has not progressed much beyond rearranging atoms on a flat surface, but in the future, it may be used to construct molecules and nano-scale machines from the bottom up.

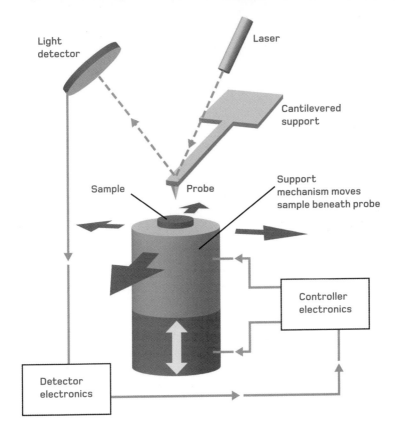

Light detector

Laser

Cantilevered support

Sample

Probe

Support mechanism moves sample beneath probe

Controller electronics

Detector electronics

Electricity

Electric charge is a fundamental property of matter that allows it to interact with electromagnetic fields. It is integral to positively charged proton particles found in the atomic nucleus, but is more commonly encountered in the form of the negative charge carried by electrons, orbiting in the outer halo of atoms or ions, or moving freely through space.

Electrical phenomena can be broadly divided into those associated with static electricity and those involving electric currents. Static electricity, as the name suggests, involves the build-up of non-moving electric charges within materials or on surfaces, the development of electric fields around and between these centres of charge, and the electric discharges (such as lightning) that ultimately occur in order to restore the balance. Materials with an accumulation of negative or net positive charge are often referred to as having negative or positive polarity. Objects of opposing polarities are attracted to one another, while those of similar polarity repel each other.

Current electricity, in contrast, involves the steady flow of electric charge through materials known as conductors. Since currents can be used to do work of various sorts, this phenomenon is far more useful: it lies at the heart of most modern electrical technology.

Although electricity involves only the flow of negatively charged particles, the movement of electrons can also be treated as a transfer of positive charge in the opposite direction. Indeed, as a result of historical accident, the conventional direction of current flow through electrical circuits is actually the way that positive charge appears to move – opposite to the direction in which electrons are actually moving. This concept of 'conventional current' is crucial to correctly understanding many electrical phenomena.

Electric charge is measured in units called coulombs, where 1 coulomb is equivalent to the charge of 6.24×10^{18} protons or electrons. A coulomb is actually defined as the amount of charge carried by a current of 1 ampere in 1 second, but as we shall see (page 206) this is a somewhat circular definition.

Static electricity

When an excess or deficiency of electrons builds up on a material without flowing away, the result is static electricity. In most circumstances, the creation of a static charge requires two materials, at least one of which is an insulator (see page 212). Electrons are transferred from one material to another when they come into contact, and the charge imbalance remains when they are removed from one another. The synthetic fibres found in various modern textiles are a familiar source of static electricity – they act as insulators and easily discard electrons to their surroundings, accumulating a net positive charge that attracts them to other surfaces that are negatively charged in comparison.

In order to even out charge between surfaces of opposite polarity, a spark of static discharge (a flow of electrons) must occur. This requires either direct contact between the surfaces, or the opening up of an electrically conductive path between them – for instance, by the ionization of air molecules.

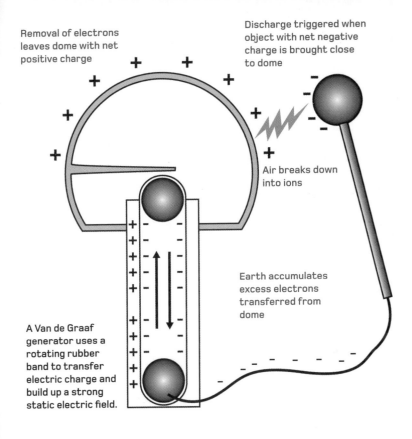

Removal of electrons leaves dome with net positive charge

Discharge triggered when object with net negative charge is brought close to dome

Air breaks down into ions

Earth accumulates excess electrons transferred from dome

A Van de Graaf generator uses a rotating rubber band to transfer electric charge and build up a strong static electric field.

Electric current

An electric current is an orderly flow of electrically charged particles through a conducting material. It usually involves the flow of negatively charged electrons, but can also involve the movement of positively charged ions in a molten material or a solution (see page 196). By convention, current is always treated as if it is the flow of (notional) positive charge. It is measured in amperes or amps, where 1 amp is equivalent to a flow of charge of 1 coulomb per second through a given surface, and is often denoted by the symbol *I*.

Because the current that a conductor can deliver is directly dependent on the number of charge carriers in a typical cross section, transmission of large currents requires physically larger wires and cables and presents a number of practical problems. The alternating current system, in which the current's direction rapidly switches back and forth, can overcome these issues (see page 230), but in most cases the behaviour of simple 'direct current' is easier to envisage.

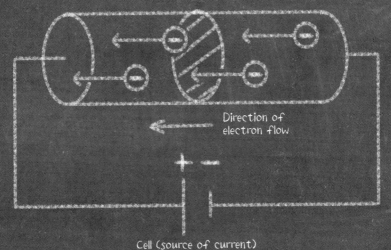

Direction of
conventional
current

Detail of
wire interior

Direction of
electron flow

Cell (source of current)

In a simple electrical circuit, electrons flow from
the negative terminal of a source of current, to the
positive terminal. Conventional current is said to
flow in the opposite direction.

Potential difference

Often known simply as voltage, potential difference is the driving force behind the flow of electric current. Just as conventional objects in a gravitational field have a measurable gravitational potential energy (see page 20) a charged particle will have a certain electrical potential energy depending on its location in an electric field. Voltage is the difference in potential energies between two points, and is defined so that the 'conventional current' (notional flow of positive charge in a conductor) always flows from higher voltages to lower ones.

The potential difference of a particular path through a conductor is defined as the energy required to move a charge from one end of the path to the other. It is measured in volts, where 1 volt is equivalent to 1 joule of energy per coulomb of charge. Moving positive charge from higher to lower voltages lessens its potential energy, but as with its gravitational analogue, it's possible to 'push' charge towards higher voltages using an energy source such as a battery.

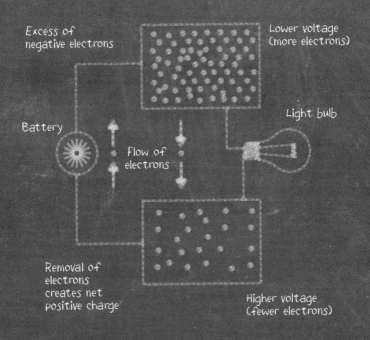

Excess of
negative electrons

Lower voltage
(more electrons)

Battery

Flow of
electrons

Light bulb

Removal of
electrons
creates net
positive charge

Higher voltage
(fewer electrons)

Electromotive force

Usually abbreviated to the initials emf, electromotive force is not in fact a force, but rather a voltage or potential difference, measured in volts. As the name implies, emf drives the motion of electric current through a conductor: the term was coined by Italian scientist Alessandro Volta, who invented the earliest form of battery – the voltaic pile – in 1800.

A wide range of different devices and phenomena can create emf within a conductor. They include the chemical energy from a battery or cell (see page 214), the influence of a changing or moving magnetic field (as in a dynamo – see page 236), thermoelectric devices that convert heat into potential difference, fuel cells and solar cells. As a result, emf can also be measured in several ways – as the simple voltage difference across the open terminals of a battery or generator, or as the work done in moving charge through a closed loop in a magnetic field.

A galvanic cell generates emf through a chemical reaction. Electrodes are pulled from the anode and travel along a wire to reach the cathode.

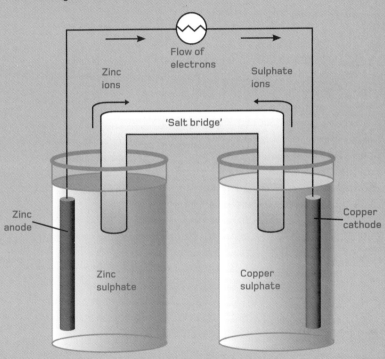

Flow of electrons

Zinc ions

Sulphate ions

'Salt bridge'

Zinc anode

Copper cathode

Zinc sulphate

Copper sulphate

Resistance and Ohm's law

When charge-carrying particles such as electrons move through a conductor, they inevitably encounter some kind of opposition to their passage thanks to interactions with the material's own atoms. This phenomenon is called resistance, and tends to dissipate energy in the form of heat. The resistance (R) of a conductor is defined by Ohm's law: $R = V/I$, where V is the potential difference across the conductor in volts, and I the current flow in amps. Resistance is measured in ohms (Ω), while conductance (G) which is simply the inverse property, is measured in siemens.

The resistance of any conductor depends on the 'resistivity' of the material it is made from, and its shape – in particular, the surface area through which current can pass. It naturally increases with temperature in most material (though see also page 240). Resistors and insulators are electrical components with deliberately high resistance that allows them to impede the flow of current and isolate conducting elements.

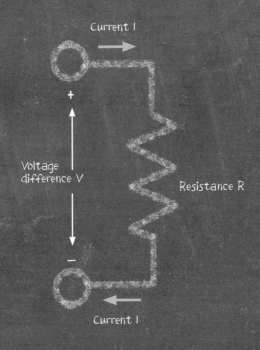

Cells and batteries

An electrochemical cell is a device that uses chemical reactions in order to generate electrical energy, and a battery is a group of linked cells. The simplest cell design uses two electrodes separated by an electrolyte (a substance that dissolves to release electrically charged ions). An electrode and its electrolyte are sometimes termed a 'half-cell' – in some designs two half-cells have different electrolytes and are connected by a 'bridge' of conducting material.

When the two electrodes are connected through an external circuit, charge is free to flow through the electrolyte and a chemical reaction begins. Negatively charged ions in one half-cell migrate towards their electrode (called the anode) where they lose electrons to become electrically neutral, while positively charged ions in the other half-cell migrate to the other electrode (the cathode) where they pick up electrons. Each half-cell creates its own emf (see page 210), and the combination of the two defines the cell's overall voltage.

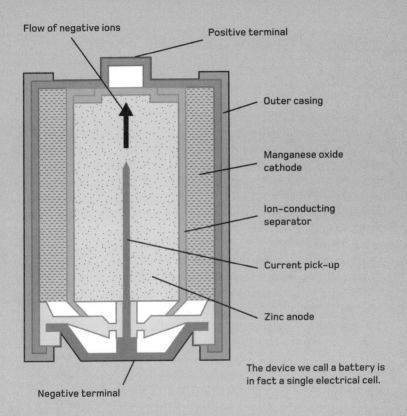

Flow of negative ions

Positive terminal

Outer casing

Manganese oxide cathode

Ion-conducting separator

Current pick-up

Zinc anode

Negative terminal

The device we call a battery is in fact a single electrical cell.

Capacitors

In order to take full advantage of the properties of electricity, it is often useful to be able to store it for later use. Capacitors offer a way of storing electrical energy within a circuit as a static electric field. They work by creating potential difference between two conductors (often called 'plates') separated by an insulating material called a dielectric. When current flows in the circuit, it cannot cross the dielectric, so equal but opposite electrical charges build up on the opposing plates, creating an electric field across the capacitor. Capacitance is measured in farads, and calculated from the simple equation $C = Q/V$ (where Q is the current flowing and V the potential difference). The larger the area of the plates, and the smaller the gap between them, the higher the capacitance. A capacitor can be discharged by bridging the plates with another conductor, but due to the high voltages they can build up, they are often drained with highly resistive materials for safety reasons. As well as simple storage, capacitors are widely used for filtering unwanted currents and smoothing current flow.

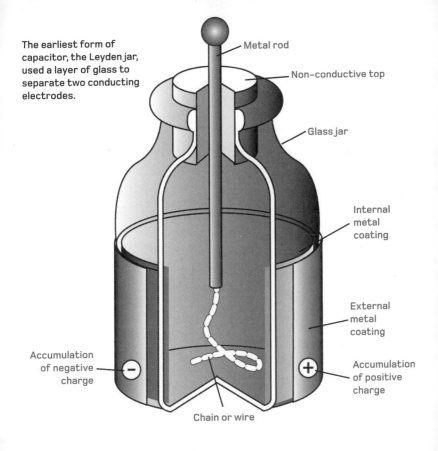

The earliest form of capacitor, the Leyden jar, used a layer of glass to separate two conducting electrodes.

Metal rod

Non-conductive top

Glass jar

Internal metal coating

External metal coating

Accumulation of negative charge

Accumulation of positive charge

Chain or wire

Circuits

A circuit is a system of electronic components connected through a loop of conductive wire. Using components such as resistors, transistors, capacitors and diodes, a circuit can perform complex tasks such as creating, receiving or amplifying electrical signals of specific frequencies, and performing calculations and measurements.

Within a circuit, components can be connected in either series or parallel arrangements. Within a series circuit, current is passed from one component to another along a single connecting wire (and a failure in a single component breaks the entire circuit). The current through each component is the same, and the voltage across the entire circuit is the sum of the voltages across indiviual components. In a parallel circuit, current flows along separate parallel wires to reach the components. The voltage across each parallel component is identical, and the current is the sum of the currents passing through each component.

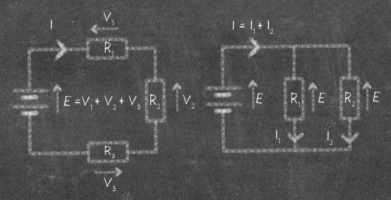

In a series circuit, the same current I flows through all components, and the overall supply voltage E is balanced by the voltages V_1, V_2, V_3 across the resistors.

In a parallel circuit, all branches are subject to the same supply voltage E, but the current is split into separate components I_1, and I_2.

Magnetism

Intrinsically linked to electricity, magnetism is an aspect of the electromagnetic force (see page 334) that causes materials to exhibit magnetic polarities (termed north or south, by analogy to Earth's own magnetic poles) that exert forces when they are brought into contact with one another. As with electrical charges, like poles repel each other while opposites attract.

Magnetism is perhaps most similar to static electricity, since magnetic materials can generate electromagnetic fields that influence their surroundings and magnetism can be transferred between materials, but there is no magnetic equivalent of flowing electric current. However, while electric charge is a fundamental property of subatomic particles, magnetic phenomena is not so elementary – it is always a result of electrical charges in motion. Hence, while many subatomic particles display an intrinsic property known as a magnetic moment, this is in fact a product of their electric charge and angular momentum or 'spin' (see page 308).

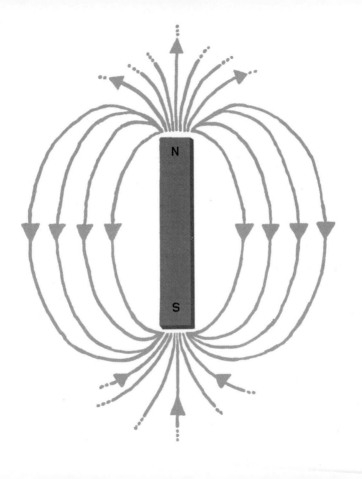

Types of magnetism

The magnetic behaviour of materials can be broadly divided into three classes – ferromagnetism, paramagnetism and diamagnetism. The weakest of these is diamagnetism, which is simply the natural tendency for all material to develop their own magnetic polarity in opposition to an applied external magnetic field. This happens as a natural result of the field's influence on electrons within the material, and means that diamagnetic objects experience a weak repulsion from magnetic fields.

Paramagnetism is a stronger form of magnetism that overwhelms diamagnetism in suceptible materials. It requires that a material has lone, unpaired electrons within its orbital shells (because the magnetic fields around paired electrons tend to naturally oppose each other and cancel out in accordance with the Pauli exclusion principle of quantum mechanics – see page 310). When an external magnetic field is applied to a paramagnetic substance, the innate 'magnetic moments' of its unpaired electrons (see page 220) will tend to line up with the field, reinforcing it and temporarily magnetizing the material. However, the magnetic moments rapidly return to their previous random alignments when the field is removed.

Ferromagnetism is the strongest form of all, and the only one in which materials can become permanently magnetized. Like paramagnetism, it requires a substance that has unpaired electrons whose magnetic moments responds to an applied field. However, ferromagnetic materials, such as iron, nickel and cobalt, also show a natural preference for having their individual magnetic moments aligned in parallel to one another (in order to preserve a stable configuration with the lowest possible energy). As a result, the magnetic moments of a ferromagnet remain aligned even after external magnetic fields are removed. This allows them to produce fields of their own. Ferromagnetism can only be 'wiped' from a material by the application of another, differently aligned field, or by heating the substance above its 'Curie temperature'. A weaker form of permanent magnetism, called ferrimagnetism, also exists in substances such as the mineral magnetite – its effects are similar to ferromagnetism, but its cause is somewhat different.

Coulomb's law and Ampère's force law

Early electrical experimenters in the 18th century observed that electrical charges close to one another experienced a force of repulsion or attraction (depending on whether they are alike or unalike), which we now understand to be caused by magnetic fields around them. In 1785, French physicist Charles-Augustin de Coulomb quantified this force in the equation:

$$F = k_e \frac{q_1 q_2}{r^2}$$

Here, q_1 and q_2 are charges, r is the distance between them, and k_e is a constant related to the electrical 'permittivity' of the surrounding material (note the similarity to Newton's law of universal gravitation – see page 32). In the 1820s, another Frenchman, André-Marie Ampère, found the force between straight, parallel conductors carrying currents I_1 and I_2 to be

$$F = 2k_A \frac{I_1 I_2}{r}$$

where k_A is the magnetic force constant, defined as 2×10^{-7} newtons/ampere^2.

Coulomb's law

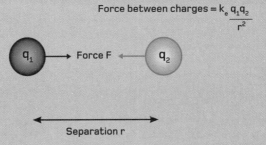

Force between charges $= k_e \dfrac{q_1 q_2}{r^2}$

q_1 → Force F ← q_2

Separation r

Ampère's force law

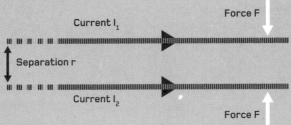

Current I_1

Force F

Separation r

Current I_2

Force F

Force between wires $= 2k_A \dfrac{I_1 I_2}{r}$

Electromagnetic induction

When a conductor is subjected to a varying magnetic field (either one that is moving, or one that is changing its strength over time), it develops an electromotive force (usually accompanied by a flow of electric current) due to a phenomenon called electromagnetic induction. The effect was discovered by British physicist Michael Faraday and the American Joseph Henry in the early 1830s, and lies at the heart of many important electrical devices including generators, motors and transformers.

Induction is caused by the magnetic field's influence on the magnetic moments of individual electrons in the conductor. Faraday later developed a law showing that the induced emf is proportional to the rate of change in the strength of the magnetic field passing through the circuit. However, the situation is complicated by the fact that induced electric currents themselves give rise to magnetic fields that oppose the field that created them (an effect known as Lenz's law).

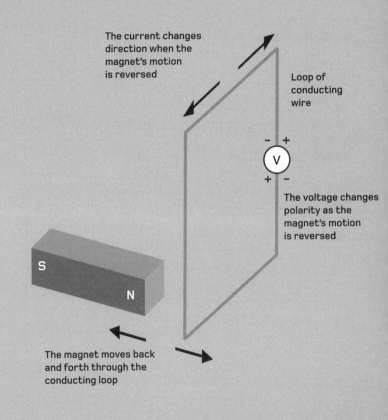

The current changes direction when the magnet's motion is reversed

Loop of conducting wire

The voltage changes polarity as the magnet's motion is reversed

S

N

The magnet moves back and forth through the conducting loop

Electromagnet

Just as a moving or changing electric field generates potential difference and causes current to flow through nearby conductors, so moving charge in a conductor creates a magnetic field around it. This phenomenon was discovered by Danish scientist Hans Christian Ørsted in 1819, when he noticed how the needle of a magnetic compass was deflected close to a wire carrying current.

Electromagnets are a common way of harnessing this effect. At their simplest, they consist of a coil containing several loops of wire, often wrapped around an insulating core for convenience. When the core is replaced with one made from a ferromagnetic material (usually iron), the effect is magnified hugely as the magnetic moments within the core (see page 220) become strongly aligned with the field from the coil. Unlike a permanent magnet, the strength of an electromagnet can be manipulated easily by altering the flow of current or switching it off completely.

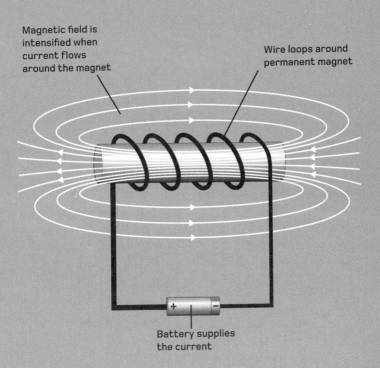

Magnetic field is intensified when current flows around the magnet

Wire loops around permanent magnet

Battery supplies the current

Alternating current

In contrast to direct current (DC), alternating current (AC) changes its direction many times every second, with the net current at any point constantly varying as a result. AC is usually described in terms of its frequency (50 or 60 Hertz in commercial power networks), its peak current and its voltage.

Intuitively, DC might seem like the most effective way of using electricity, but there are few situations where its properties are crucial. AC's main advantage, meanwhile, is that its voltage can easily be altered using a transformer (see page 232). High voltages are ideal for transferring energy over long distances (since large amounts of energy can be tranferred using only small currents, and it is the level of current that governs the amount of energy lost to resistance). However, high currents are ideal for doing actual electrical work in domestic and industrial situations. AC is relatively easily rectified to produce DC current where needed – another reason why it has become the standard means of electrical power distribution.

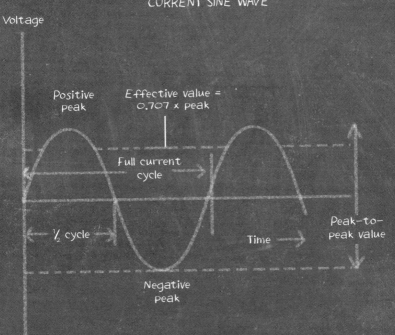

FEATURES OF AN ALTERNATING
CURRENT SINE WAVE

Voltage

Positive
peak

*Effective value =
0.707 x peak*

Full current
cycle

½ cycle

Time ⟶

Peak-to-
peak value

Negative
peak

Transformer

A transformer is a device that transfers electrical energy between two conductors using the principles of induction (see page 226). It uses two conducting coils (the primary and secondary windings) wrapped around a ferromagnetic core. When current flows through the primary winding, a magnetic field is created in the core, which in turn induces a potential difference and flow of current in the secondary winding. The induction is driven by the *changing* strength of the current and magnetic field, so transformers only work efficiently with AC electricity, in which the current is constantly varying.

The most important property of any transformer is that the potential difference V (and therefore the flow of current I) in the secondary coil is governed by the ratio of the number of turns in the primary and secondary windings, N_p and N_s. This gives rise to the transformer equation:

$$V_p / V_s = N_p / N_s = I_s / I_p$$

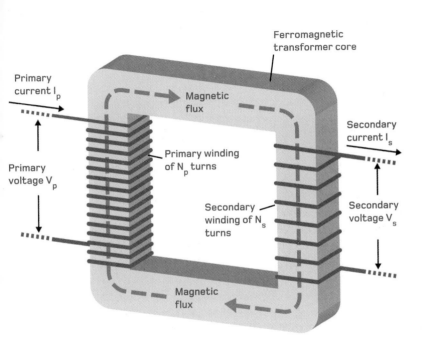

Electric motor

An electric motor is a device that uses electromagnetic effects to create mechanical motion. In general, motors rely on the forces generated between conductors (see page 224). They consist of two key elements – a 'rotor' that generates the mechanical motion and a 'stator' that usually produces the magnetic fields needed to drive the motor. At least one element must incorporate 'windings' – conducting coils that generate a magnetic field when current passes through them. The other may use a permanent magnet or an electromagnet.

The principle behind the motor is simple: changing current in the windings produces a changing magnetic field that drives a repulsion from the other magnetized element and causes the rotor to rotate slightly. The challenge lies in making the rotation continuous: DC motors use a switch called a commutator to reverse the current direction with every half-turn of the rotor (so the magnetic force always pushes in the same direction) while AC motors make use of the current's natural reversals.

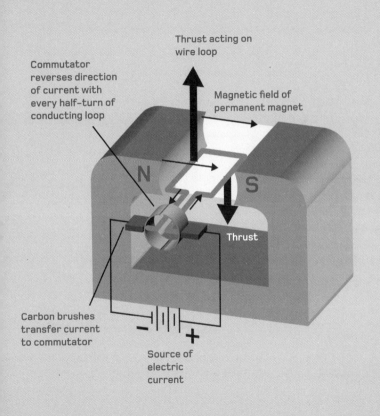

Thrust acting on
wire loop

Commutator
reverses direction
of current with
every half-turn of
conducting loop

Magnetic field of
permanent magnet

N

S

Thrust

Carbon brushes
transfer current
to commutator

Source of
electric
current

Generator

A generator is a device that harnesses mechanical movement and electromagnetic induction (see page 225) in order to produce electric current – as such, it is more or less the opposite of an electric motor. Generator designs are varied, but many use the dynamo principle – mechanical power in some form drives the rotation of a drum wrapped with coils of conducting wire, mounted in the middle of a fixed ring of permanent magnets. As the coils move through the magnetic field, a potential difference is induced within them, driving the flow of electric current. Power sources for this kind of generator range from wind and water pressure to pedal power, and from steam turbines to internal combustion engines.

As their wire coils rotate, dynamos naturally produce AC current – indeed, if used as a raw power supply they are often called alternators. However, in applications such as the bicycle dynamo, a device called a commutator (see page 234) is used to ensure direct current, flowing steadily in one direction.

Analogue and digital electronics

So far in this section, we've assumed that electric current varies continuously – in other words, it can take on any value within a certain range. This 'analogue' behaviour makes continuously varying current useful for representing other similarly variable properties. Traditional radio systems, for example, encode sound into a varying current, which is then used to broadcast signals that can be decoded by a receiver.

But analogue systems have some problems. Most significantly, they are prone to interference or 'noise' – random fluctuations introduced as the signal moves through the system, which are

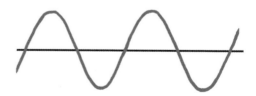

An analogue signal such as an electric current can vary continuously and represent any value between its extremes.

difficult to remove and degrade the quality of information that can be extracted at the other end of the process.

The alternative to an analogue signal is a digital one, in which current varies in sharply defined steps. In practice, digital is synonymous with the binary system, in which current has just two values (present or absent), indicating the digits one and zero. The binary or base-2 number system is a 'place-value' system similar to our decimal one, but with just two digits, so that, for example, the numbers 0, 1, 2, 3 and 4 become 0, 1, 10, 11 and 100. Transforming an analogue signal into digital involves repeatedly measuring or 'sampling' its value, and converting the value at a particular instant into binary. The result is a stream of 1s and 0s that is largely immune to the problems of signal degradation, and is also ideal for manipulation by electronic components such as diodes (see page 242).

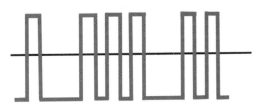

A digital signal has only two values, representing binary 1s and 0s.

Semiconductor

A semiconductor is a material with unique electrical properties that, as the name suggests, place it midway between conductors and insulators. Semiconductor materials are typically compounds of elements from Group 14 of the periodic table (see page 184), such as silicon and germanium. The crystalline lattice they form as pure solids tends to have either a natural excess of negatively charged electrons (in an n-type semiconductor) or distinct 'holes' with net positive charge (in a p-type semiconductor), either of which may act as charge carriers. P-type and n-type regions can be created with relative ease by 'doping' parts of a semiconductor crystal with impurities.

Semiconductors can display a wide range of useful properties, such as the one-way flow of charge and increased conductivity with higher temperatures. They are engineered to create a huge variety of electrical components, such as diodes and transistors, and form the foundation of modern electronics.

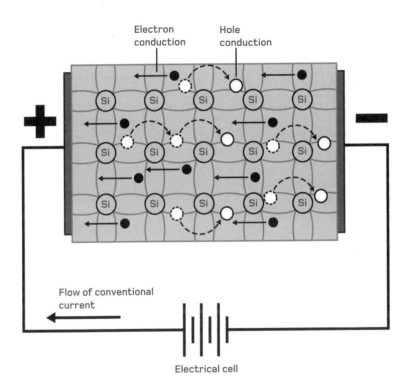

Diode

A diode is a relatively simple electrical component that acts rather like a valve in a physical system, allowing current between two terminals to flow freely in one direction, but strongly resisting (if not entirely blocking) its passage in the opposite direction. Indeed, many early diode designs in the era before miniaturization were known as valves. The most successful of these, the thermionic valve, is illustrated opposite. In semiconductor materials (see page 240), diodes can be produced by creating a 'p-n junction' with an excess of positive 'holes' on one side and negative electrons on the other. Current will pass from the n-type, electron-enriched side to the p-type side, but will not flow back in the opposite direction.

Diodes can perform many different functions – one of the most useful is rectification of alternating current to produce direct current (see page 230). It's also possible to make diodes that will only allow current to flow in the 'forward' direction if a certain potential difference threshold is overcome.

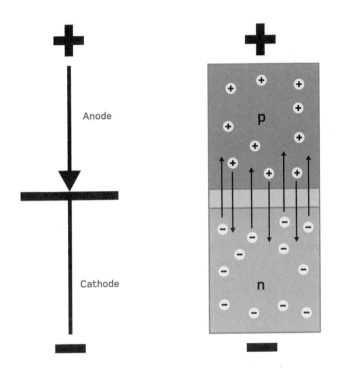

The conventional representation of a forward-biased p-n junction diode (left) and a schematic (right) showing the flow of charge within it.

Transistor

The transistor forms the basis for a huge range of modern electronic devices. It has at least three terminals, designed so that the voltage or current applied across one pair controls the flow of current across the other pair. A transistor can therefore act as a switch (with the output turned on or off by the applied control current), or as an amplifier, since the output can mirror the control signal but provide an arbitrarily higher current. Various arrangements of transistors can be used to perform logic operations (see page 246).

Transistors began life as cumbersome devices based on glass valves kbiwb as thermionic triodes, but today are commonly made from semiconductor materials (see page 240). The most widespread designs are the bipolar junction transistor (a 'semiconductor sandwich' with either an 'npn' or 'pnp' structure), and the field effect transistor, a four-terminal design in which the flow of electricity between two terminals is controlled by the electric field between the other two.

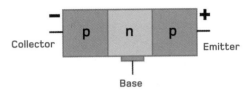

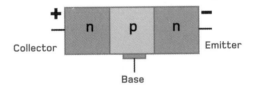

A bipolar junction transistor comprises a layer of either p-type or n-type semiconductor (the base) sandwiched between two layers of the opposite type (the emitter and collector). A large current flowing between the two sides of the sandwich can be regulated by a small current applied to the base.

Logic gates

By arranging a number of electronic components such as transistors and diodes (see pages 244 and 242) it is possible to build devices that perform logic operations – for instance analysing two signals and only generating an output if both have a certain value (an 'AND' operation). Discrete units of this kind, known as logic gates, lie at the heart of computer technology.

The principal types of logic gate are the AND, the OR (in which a signal at either of the two input terminals produces an output) and the NOT (which produces an output only when there is no input). Combinations of these basic types can be used to construct gates capable of performing more complex logical operations such as NAND and NOR (a complete list of basic logic operations is given opposite). Early electronic systems used large evacuated glass valves and transistors, but in modern computers they usually take the form of microscopic semiconductor components on integrated circuits (see page 248).

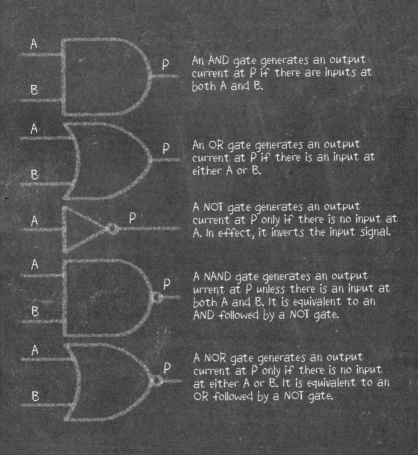

An AND gate generates an output current at P if there are inputs at both A and B.

An OR gate generates an output current at P if there is an input at either A or B.

A NOT gate generates an output current at P only if there is no input at A. In effect, it inverts the input signal.

A NAND gate generates an output urrent at P unless there is an input at both A and B. It is equivalent to an AND followed by a NOT gate.

A NOR gate generates an output current at P only if there is no input at either A or B. It is equivalent to an OR followed by a NOT gate.

Integrated circuit

An integrated circuit (IC) is a small block of semiconductor material (see page 240) with millions or even billions of separate electronic components etched into its surface. Ubiquitous in modern technology, its functions range from data storage to complex computer processing tasks. Most modern integrated circuits use CMOS (complementary metal-oxide semiconductor) technology in which different areas of the semiconductor are 'doped' with different metal oxides in order to give them p-type or n-type properties. These junctions build up diodes and transistors, logic gates and more complex structures. Different parts of the substrate (usually a wafer of pure silicon) are imprinted with either type of semiconductor and fine lines of insulating material or conducting metal such as aluminium, using a technique called photolithography, analogous to traditional photography. At the very smallest sizes, circuits can respond to the flow of individual electrons, although such highly specialized systems require extremely low temperatures in order to operate.

Superconductivity

The passage of electrons through a conducting lattice almost always encounters some form of resistance, even in the best conductors, and the effect usually increases with temperature. Semiconductors, in contrast, produce lower resistance at higher temperatures. However, superconductors are a rare class of materials that, in the right conditions, can conduct electricity perfectly, with zero resistance. Superconductors not only act as perfect electrical conductors, but also actively expel magnetic fields passing through them.

A wide range of materials, including some metals, alloys, ceramics and even organic (carbon-based) molecules, display superconductivity when they are cooled below a certain critical temperature. In most cases, this temperature is within 30°C (54°F) of absolute zero (see page 108), but in so-called 'high-temperature' superconductors, it is significantly higher (though still very cold by everyday standards).

Superconductivity seems to arise in a variety of different ways, but the most common form is explained by the BCS (Bardeen-Cooper-Schrieffer) theory. Developed in the 1950s, this theory involves electrons binding together to form weakly bonded 'Cooper pairs' at extremely low temperatures. This pairing allows them to behave in a similar way to the class of quantum particles known as bosons (see page 310). In many ways, it seems, BCS superconductivity is similar to the quantum phenomenon of superfluidity (see page 320). The Cooper pairs are all able to take on precisely identical properties to one another, and this dramatically reduces their interactions with the surrounding material.

Traditional superconductivity has many applications, including production of the powerful superconducting electromagnets used in MRI scanners (see page 312) and particle accelerators (see page 326). The phenomenon can also be put to work in delicate sensors for measuring heat, light and magnetic fields. High-temperature superconductors, meanwhile, currently remain limited to the distinctly chilly temperatures below -140°C (-220°F). If more practical examples can be found, they have the potential to revolutionize many aspects of everyday technology.

Photomultiplier tube

A photomultiplier tube (PMT) is a specialized vacuum tube that converts light into electronic signals. It relies on the photoelectric effect – the emission of electrons when a material is struck by photons of light, as discovered by Heinrich Hertz in 1897 and explained by Einstein in 1905 (see page 286).

At the front of the tube, photons of light strike a negatively charged surface called the photocathode, causing it to emit electrons. These are then focused into a beam by a focusing electrode, before bouncing between a series of electrodes called dynodes, each held at a higher positive voltage than the last. The electric field between the dynodes creates a cascade effect, releasing more electrons at each stage until they reach the final electrode (the anode) and generate a significant current. PMTs formed the basis for the first electronic television cameras and, although they have been supplanted in that area by CCDs (see page 256), they still have a wide range of uses in science, industry and medicine.

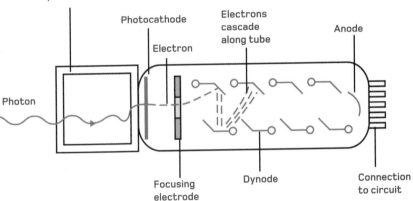

Scintillator material
fluoresces when
excited by incoming
photon

Photocathode

Electron

Electrons
cascade
along tube

Anode

Photon

Focusing
electrode

Dynode

Connection
to circuit

Cathode ray tube

Not all electronic devices involve electrons moving through conducting wires – the cathode ray tube (CRT) controls beams of electrons moving through an evacuated glass chamber. The electrons emerge from a device called an electron gun and are accelerated and focused into a tight beam by electric fields. This beam is then deflected by further magnetic fields in order to trace patterns on a screen at the other end of the tube. A fluorescent coating on the inner surface of the screen glows where the beam strikes it, making these patterns visible. CRT technology forms the basis of oscilloscopes and other display devices, and was for a long time the preferred means of displaying television pictures.

The electron gun itself is a negatively charged electrode or cathode that is heated until electrons break free from its surface (a process called thermionic emission). British physicist J.J. Thomson's investigation of these 'cathode rays' led directly to his discovery of electrons in 1897.

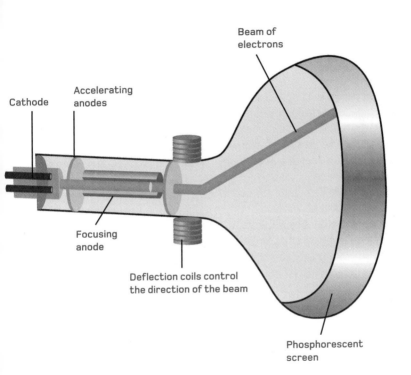

Cathode

Accelerating anodes

Focusing anode

Deflection coils control the direction of the beam

Beam of electrons

Phosphorescent screen

Charge-coupled device

Better known by the simple acronym CCD, a charge-coupled device is a specialized type of integrated circuit found in many modern electronic imaging systems. It converts photons of light striking an array of picture elements, or 'pixels', each a few microns (millionths of a metre) across into electric charge.

Each pixel is a capacitor (see page 236) in which a thin layer of insulating silicon dioxide acts as a dielectric material between the semiconductor substrate and an overlying conductor. Photons striking the upper surface generate electrons through the photoelectric effect (see page 286), and the charge builds up across the array in strict proportion to the number of photons striking different areas, forming 'potential wells'. This makes CCDs a very efficient means of collecting light. At the end of the image exposure, the first pixel in the array passes its charge to an amplifier that converts it to a voltage, while each remaining pixel transfers its charge to its neighbour. The process repeats until the entire array has been read.

Anatomy of a single CCD pixel

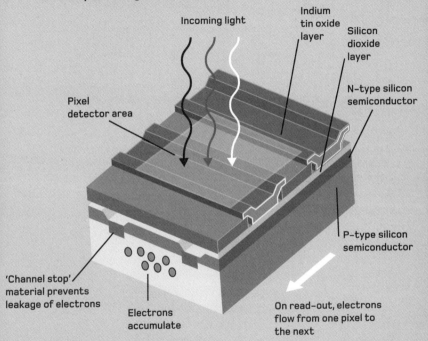

Incoming light

Indium tin oxide layer

Silicon dioxide layer

N-type silicon semiconductor

Pixel detector area

P-type silicon semiconductor

'Channel stop' material prevents leakage of electrons

Electrons accumulate

On read-out, electrons flow from one pixel to the next

Nuclear physics

Until the early 1900s, physicists assumed that internally, atoms was more or less uniform. Electrons, the only known subatomic particles, were seen as small negatively charged bodies floating in a sea of overall positive charge. It was only in 1911 that New Zealand-born scientist Ernest Rutherford discovered that most of an atom's mass was concentrated in a tiny region at the centre, and nuclear physics was born.

Today we know that a typical atomic nucleus is just a few femtometres (millionths of a nanometre) across, and that nuclei consist of two types of heavy particle — positively charged protons and uncharged neutrons. The development of nuclear physics has largely been concerned with the relationships between these two particles, and the nuclear transformations associated with radioactivity. The first radioactive materials were discovered in the late 1890s, and were soon recognized as both an important tool for learning more about atomic structure, and a potentially powerful source of energy.

The natural radioactive processes encountered on Earth all involve the splitting apart or 'decay' of a heavy, unstable isotope (see page 170) into lighter and potentially more stable ones. Physicists recognize three common tyes of radioactive emission associated with decay: in increasing order of their penetrating ability these are known as alpha particles, beta particles and gamma rays. Natural decay processes can be used as a 'radiometric clock' with a wide range of uses, while the deliberate splitting or fission of heavy particles can be a powerful source of energy for use in both power generation and the simplest atomic weapons.

An even more powerful nuclear process, known as fusion, does not occur naturally on Earth, but is widespread in the Universe, since it is the mechanism by which the stars shine. Fusion requires extreme temperatures and pressures – in the centre of our own Sun, it occurs at temperatures of around 15 million °C (27 million °F) and densities 160 times greater than that of water. In such conditions, the nuclei of light elements are forced together to create heavier ones, releasing large amounts of excess energy in the process. However, human exploitation of fusion power is still limited to experimental reactors (see page 280) and the hydrogen bomb.

Alpha decay

The radioactive process known as alpha decay involves the emission of a heavy particle known as an alpha particle, containing two protons and two neutrons. Alpha particles are identical to a nucleus of helium-4, the second lightest element in the Universe. Following alpha emission, the atomic number of the nucleus is reduced by two, and its atomic mass by four.

Alpha decay is the most common means for unstable nuclei to shed excess protons and neutrons (collectively known as nucleons). It can be triggered by bombarding elements with atomic numbers higher than 52 (tellurium) with alpha particles, and occurs spontaneously in elements with atomic numbers above 90 (thorium). However, it is not the only way in which elements can shed heavy particles – other rarer forms of nucleon emission include the direct expulsion of a single proton or neutron, or a pair of protons. In certain situations the nucleus can also split to release atomic nuclei that are heavier and more complex than helium.

A typical alpha decay process

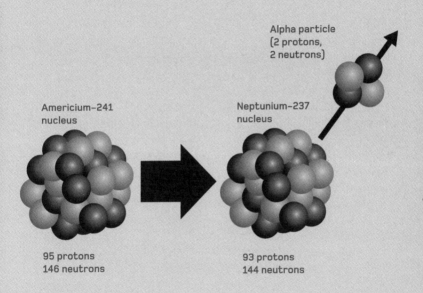

Alpha particle
(2 protons,
2 neutrons)

Americium-241
nucleus

Neptunium-237
nucleus

95 protons
146 neutrons

93 protons
144 neutrons

Beta decay

Beta decay is a radioactive process that involves the release of a lightweight particle (usually an electron but sometimes its antimatter counterpart, a positron – see page 168) directly from the nucleus. This raises an obvious question: based on our understanding of atomic structure (see page 160) – what is an electron doing in the nucleus in the first place?

The answer, it seems, is that the electron simply doesn't exist prior to the decay event. The driving force behind beta decay is actually the spontaneous transformation of a neutron into a proton or (less commonly) a proton into a neutron, in order to make the nucleus more stable and evenly balanced between particle types. The electron or positron is simply created to balance out the negative or positive charge that is 'shed' by the proton/neutron transition. As a result, in standard beta decay, the decayed nucleus has more or less the same atomic mass as its progenitor, but an atomic number increased by one.

Beta-minus decay process

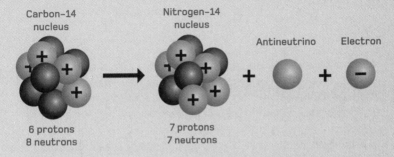

Carbon-14
nucleus

Nitrogen-14
nucleus

Antineutrino

Electron

6 protons
8 neutrons

7 protons
7 neutrons

Beta-plus decay process

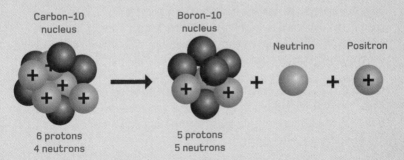

Carbon-10
nucleus

Boron-10
nucleus

Neutrino

Positron

6 protons
4 neutrons

5 protons
5 neutrons

Gamma emission

In contrast to the particles released by alpha and beta decay, gamma decay releases only electromagnetic radiation, in the form of high-energy gamma rays (see page 138). Other forms of fission often leave the nucleus in an excited state, and as the nucleons rearrange themselves into the configuration with the lowest energy, any excess can be released as a burst of electromagnetic rays. Gamma radiation is also the principal means by which nuclear fusion reactions release excess energy. The density of material in the hearts of stars like the Sun ensures that most of the gamma-ray energy is scattered (see page 90) and heats the solar interior, so that by the time solar radiation escapes from the surface of the Sun, it has mostly been diminished to less energetic forms.

Gamma radiation is by far the most penetrating and harmful form of radioactive emission, and can only be blocked effectively using thick barriers of dense material, such as lead.

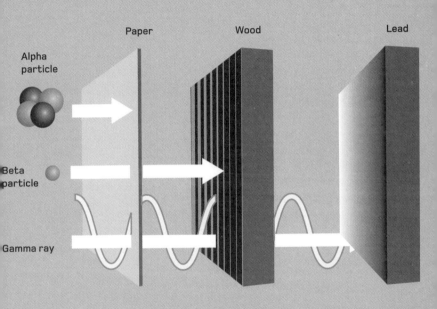

Alpha
particle

Beta
particle

Gamma ray

Paper

Wood

Lead

Penetrating power of different
radioactive emissions

Neutrino

A neutrino is an electrically neutral, near-massless particle created as a by-product of many nuclear reactions. Its existence was suggested by Austrian theoretical physicist Wolfgang Pauli in 1930 as a possible means of conserving the quantum 'spin' property (see page 308) during beta decay.

Neutrinos interact only weakly with other forms of matter and are notoriously hard to detect. Trillions of them (mostly generated by the Sun at speeds close to that of light) pass through our bodies every second, and they pass through solid rock just as easily. For this reason, neutrino detectors (often looking for Cherenkov radiation – see page 150) are placed deep underground, beyond the reach of other particles. For a long time, neutrinos were though to be massless, but we now know that they oscillate between three different forms – a process that requires some small amount of mass. As a result, they may make some contribution to the problem of so-called 'dark matter' (see page 402) in the Universe.

Pattern of Cherenkov radiation generated by a neutrino passing through an underground detector.

Half–life

The unpredictable nature of the processes involved in radioactive decay renders individual decay events impossible to predict – these events rely on a particular reconfiguration of the atomic nucleus that allows emitted particles to overcome the forces binding them together with the rest of the nucleus (see pages 342 and 346). Despite this, physicists can predict the properties of radioactive isotopes en masse, through a property known as the half-life.

Statistically speaking, half-life is the time required for any quantity to diminish to half of its original value – in terms of radioisotopes this is the time taken for half of the isotopes in a sample to decay, halving the sample's overall radioactivity if (and only if) its decay products are non-radioactive. Half-life is usually denoted $t_{1/2}$. It allows the calculation of two other important properties: the mean lifetime (τ – the average lifetime of a radioisotope before decay) and the decay constant (λ – the rate of decay per unit of time, given by $1/\tau$).

The decay curve of a typical radioactive isotope begins with a large population of the original 'parent' isotope, which decay at random over time to produce 'daughter' isotopes.

Parent isotope

Daughter isotope

Percentage of parent isotope remaining

100

50

25

12.5
6.25

0   1   2   3   4

Time (half-lives) —>

Binding energy

Nuclear binding energy is a measure of the quantity of energy required to reduce a nucleus into its component protons and neutrons. It reveals itself as a small but significant difference or 'mass deficit' between the mass of a nucleus and the mass of its component nucleons. The mass of the nucleus is always less than that of the nucleons, and this is described as a positive binding energy – a small amount of mass has been converted directly into energy in accordance with Einstein's famous equation $E = mc^2$.

Every element and isotope has its own unique binding energy, allowing the construction of a 'binding energy curve' as shown opposite. For lightweight elements, binding energy increases with mass, but fusion of light elements releases energy because the combined binding energy of *all* the nuclei involved is reduced in the process. For elements above atomic number 26 (iron), binding energy decreases with increasing mass, and so nuclear fission releases energy.

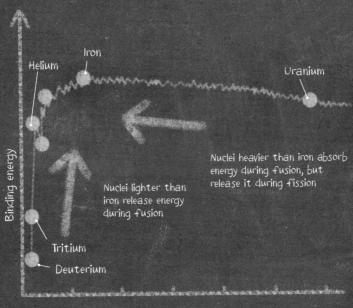

Decay series

The decay of unstable radioisotopes through nuclear fission often gives rise to a new generation of isotopes that may be just as unstable, if not more so. As a result, reaching a final stable state often involves a number of separate decay events, collectively known as a decay series or chain. The decay series of uranium-238 (^{238}U), for example, takes 19 separate steps to eventually produce the stable lead isotope ^{206}Pb.

In any individual stage of a decay chain, a parent isotope is said to give rise to a daughter isotope. When handling radioactive materials it is often the presence of daughter isotopes that creates problems – for instance, isotopes of the commonly used nuclear fuel uranium have extremely long half-lives, and so uranium itself is not significantly radioactive in quantity. However, the products of its decay series include isotopes of radium and radon, which both have significantly shorter half-lives, and it is the radioactivity from these elements that makes handling and processing uranium ore dangerous.

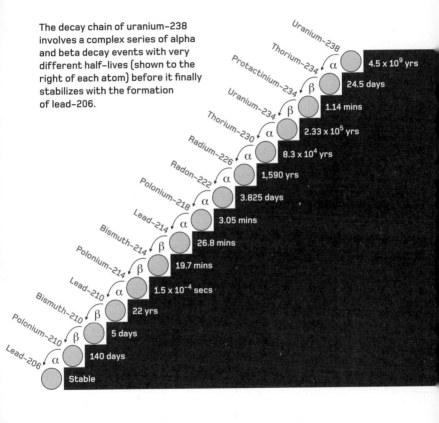

The decay chain of uranium-238 involves a complex series of alpha and beta decay events with very different half-lives (shown to the right of each atom) before it finally stabilizes with the formation of lead-206.

Uranium-238 — 4.5 × 10⁹ yrs

Thorium-234 α — 24.5 days

Protactinium-234 β — 1.14 mins

Uranium-234 β — 2.33 × 10⁵ yrs

Thorium-230 α — 8.3 × 10⁴ yrs

Radium-226 α — 1,590 yrs

Radon-222 α — 3.825 days

Polonium-218 α — 3.05 mins

Lead-214 α — 26.8 mins

Bismuth-214 β — 19.7 mins

Polonium-214 β — 1.5 × 10⁻⁴ secs

Lead-210 α — 22 yrs

Bismuth-210 β — 5 days

Polonium-210 β — 140 days

Lead-206 α — Stable

Radiometric and isotopic measurements

By measuring the proportion of different isotopes within a material (for instance, a geological sample or an archaeological find), scientists can discover a surprising amount about it. Samples are usually analysed using a mass spectrometer (see page 172) to break their components down and analyse the distribution of isotopes.

Perhaps the best-known use of isotopes is in radiometric dating to reveal the age of rocks and organic materials. The mix of isotopes in a molten rock when it solidifies, or in a living organism when it dies, reflects the environment of its time, so if we can measure the proportions of parent and daughter isotopes in the sample, we can work out how much time has passed since the 'radiometric clock' started ticking.

However, radiometric dating is not entirely straightforward – radioisotopes with particularly long half-lives tend not to yield the most accurate dates, while those with shorter half-lives may decay so rapidly that they have effectively disappeared after a relatively short period of time (for instance, carbon-14, used in the well-known technique of carbon dating, is useless for samples older than 60,000 years).

The varying weight of isotopes can create other issues with their analysis, since it may cause them to accumulate preferentially within certain environments. Carbon-14, for example, tends to collect in oceans, so the carbon dating of marine organisms can be 'skewed' compared to that of land-dwelling ones. However, such effects are not merely problems to be overcome – they also provide valuable information about past environments in themselves. For example, water molecules containing the rare heavy oxygen isotope oxygen-18 tend not to evaporate from the oceans during cold climate conditions, and so its absence from certain layers within polar ice sheets (whose water ultimately came from the atmosphere) is a useful indicator of ancient ice ages. Isotope ratios can even tell us about the present-day climate – the quantity of atmospheric carbon dioxide containing carbon-14 is crucial evidence that the increase in this greenhouse gas is due to burning of fossil fuels enriched with the heavy isotope.

Geiger counter

The Geiger counter is the most familiar instrument for detecting radioactive emission. This hand-held device emits a click each time it detects the passage of a radioactive particle and often has an electronic read-out to indicate the overall rate of emission. Most designs use a thick-walled tube with a window at one end, filled with low-pressure, inert gas. An isolated wire runs down the middle of the tube, while the inner surface of the tube is also electrically conductive, so that when the Geiger counter is switched on, a potential difference of several hundred volts is generated between them.

When an electrically charged alpha or beta particle passes through the thin window into the tube, it briefly allows the gas to conduct (in some designs, the impact of a gamma ray on the tube interior can produce the same effect). A spark then passes between the two electrodes, and an avalanche effect mutliplies the number of electrons passing across the gas to produce a measurable current through the circuit.

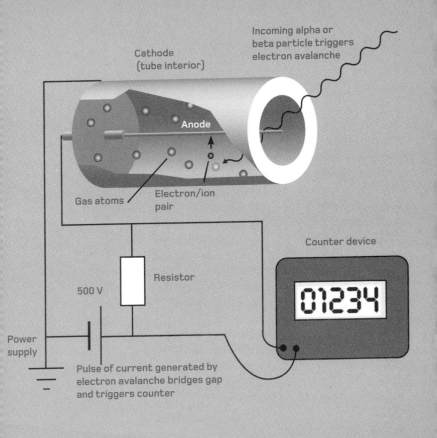

Incoming alpha or
beta particle triggers
electron avalanche

Cathode
(tube interior)

Anode

Gas atoms

Electron/ion
pair

Counter device

01234

Resistor

500 V

Power
supply

Pulse of current generated by
electron avalanche bridges gap
and triggers counter

Fission energy

In order to harness the energy of nuclear fission, physicists take advantage of a phenomenon called induced fission. While individual nuclear reactions are normally unpredictable, in the right circumstances and using the right radioisotope (most commonly uranium-235), the impact of a neutron particle can stimulate a fission event, producing two lighter nuclei and additional fast-moving lone neutrons. These neutrons can then trigger new fission events, setting off a 'chain reaction'.

Several different reactor designs have been developed, but the principal aim is always to sustain the chain reaction in a controlled and safe way, using a medium called a moderator to slow down the neutrons and regulate the chain reaction. The speed at which a reaction proceeds depends on a number of factors, including the amount of 'fissile' material present and the temperature. Energy is extracted from the reactor core by heating a 'working fluid' (usually water) and using the steam generated to drive an electric turbine.

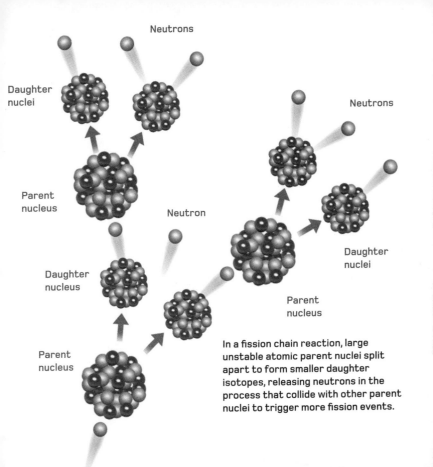

Neutrons

Daughter nuclei

Parent nucleus

Neutron

Daughter nucleus

Parent nucleus

Neutrons

Daughter nuclei

Parent nucleus

In a fission chain reaction, large unstable atomic parent nuclei split apart to form smaller daughter isotopes, releasing neutrons in the process that collide with other parent nuclei to trigger more fission events.

Fusion energy

Fusion power offers the potential for generating enormous amounts of clean, cheap energy. In essence, a fusion reactor aims to replicate events that take place in the heart of the Sun, fusing hydrogen nuclei to create the next lightest element, helium, and releasing energy. Fusion of 'normal' hydrogen nuclei requires conditions that cannot yet be achieved on Earth, so experimental reactors aim instead to replicate later stages in the solar fusion chain, combining the nuclei of the heavier hydrogen isotopes deuterium and tritium (both extracted from sea water with relative ease) to create helium. Although some radioactivity is generated, the isotopes involved have short half-lives, allowing for easier waste management.

So far, fusion has been achieved in doughnut-shaped chambers using electromagnetic fields or lasers to compress and heat the fuel. Recent experiments have approached 'break-even', at which they generate more energy than is needed to power them, but there are still many technical hurdles to overcome.

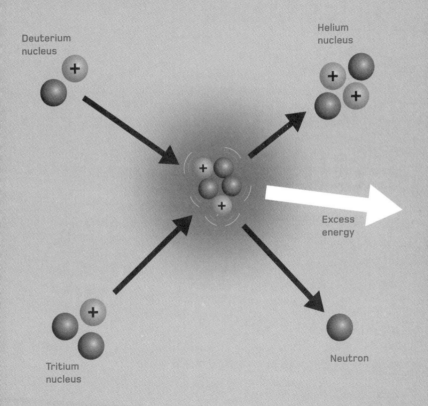

Deuterium
nucleus

Helium
nucleus

+

+

+

+

+

Excess
energy

Tritium
nucleus

Neutron

Nuclear weapons

A nuclear bomb generates an explosion by triggering a violent runaway chain reaction. The simplest design – used in the 'atomic bombs' dropped at the end of the Second World War – involves combining two masses of fissile material (usually either uranium artificially enriched with the uranium-235 isotope, or the even heavier element plutonium) to create a 'critical mass' in which a chain reaction similar to those used in controlled fission reactors is allowed to run out of control.

An even more powerful weapon, the hydrogen bomb, uses a small 'primary' fission bomb that triggers fusion in a secondary stage enriched with heavier isotopes of hydrogen. This fusion stage is in turn encased in depleted uranium (a by-product of the enrichment process), and the blast of neutrons produced by the fusion explosion can trigger fission of these atoms to release even more energy. In theory, additional fusion stages can be added to achieve staggering explosive yields, measured in megatons (equivalent to millions of tons of TNT).

Schematic of a hydrogen bomb

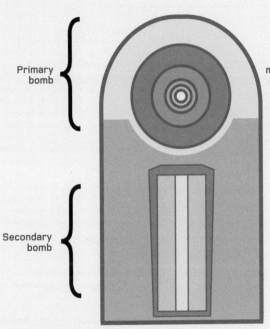

Primary bomb

The primary bomb consists of fissile materials surrounded by lenses of high explosive that force them together.

Secondary bomb

The secondary bomb consists of a plutonium 'spark plug' surrounded by lithium-6 deuteride, which supplies the fuel for the actual fusion explosion. Neutrons from the fusion trigger fission in the depleted uranium casing.

Quantum physics

When its foundations were laid in the early part of the 20th century, quantum physics offered a revelatory, and at times unnerving, new way of looking at the world. The quantum revolution seemed to sweep away the old certainties of classical physics, replacing them with talk of probabilities, superpositions of possible outcomes and, most notably of all, *uncertainty*.

Put simply, quantum physics is the physics of the very small – its effects only become noticeable at the level of individual atoms and subatomic particles, while leaving the large-scale 'macroscopic' world largely unscathed (so that the laws of classical physics are still usable in most situations). Crucial to the quantum world is the concept of wave-particle duality – the idea that waves and particles are not the distinct and very different phenomena they normally appear to be, but that instead waves (specifically electromagnetic waves) can sometimes behave like particles, and vice versa.

Another key feature of the quantum model is the idea that on very small scales, apparently continuous properties are 'quantized' (split into very small but discrete units). This first became apparent through the discovery of photons (see page 142) in the early 1900s – 'particles of light' that effectively quantize the energy delivered by electromagnetic waves. From the 1920s onwards, it became clear that, just as particles could display wave-like behaviour, many of their properties were also quantized and best described using 'quantum numbers' (see page 306). For example, the angular momentum of electrons can only display specific quantized values.

Perhaps the most challenging aspect of quantum physics, however, is the idea that the wave-like features of particles give them an inherent unpredictability – we can no longer talk in terms of certainties, only in terms of the probability of different outcomes, described by the daunting Schrödinger wavefunction. The struggle to understand the way in which the 'probabilistic' nature of the quantum world connects to the apparently absolute or 'deterministic' nature of the large-scale Universe has given rise to a number of 'interpretations' of quantum physics, often presenting not only scientific, but also philosophical challenges (see pages 298–305).

Photoelectric effect

The photoelectric effect is simply the emission of electrons from the surface of certain metals when they are exposed to light. In many ways, the effect is simple to understand – energy supplied by radiation is absorbed by electrons and allows them to break free of their orbits around the atomic nucleus. The problem for physicists studying the effect at the turn of the 20th century was that it only happened for certain frequencies of light. No matter how much red light illuminated a surface, it remained unaffected, but even weak blue light would trigger emission. Yet surely the red light could be delivering more energy?

The solution, outlined by Albert Einstein in 1905, was that the electrons were being energized not by a continuous stream of light, but by individual 'packets' or photons (see page 142) whose energy depended on their frequency or colour – intense red light merely delivers large quantities of lower-energy photons. This discovery became known as the quantization of light.

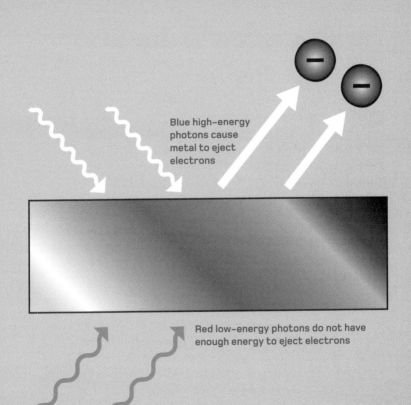

Blue high-energy photons cause metal to eject electrons

Red low-energy photons do not have enough energy to eject electrons

Wave-particle duality

Essential to quantum physics is the idea that wave-like phenomena can sometimes behave like particles, and vice versa. This concept was first applied to electromagnetic radiation, explaining not only the photoelectric effect (see page 286), but also the ability of light to travel through vacuum without a medium to carry it. The idea of 'quantized light' was implicit in a model for the characteristic patterns of black body radiation (see page 148) put forward in 1900 by German physicist Max Planck, but he treated it purely as a mathematical tool, and it was only in 1905 that Einstein suggested that photons of light might be a physical reality.

Perhaps the strangest aspect of this 'wave-particle duality' is that it is impossible to demonstrate both the wave-like and particle-like behaviours of light (or matter, since duality cuts both ways – see page 290) at the same time. A popular demonstration of this 'quantum weirdness' is the 'dual slit' experiment illustrated opposite.

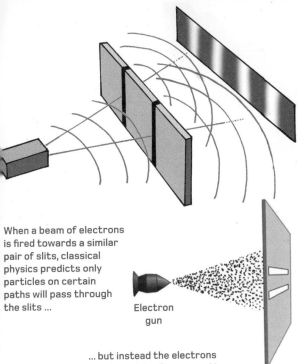

Light beams passing through two narrow slits diffract and interfere with each other, forming interference patterns that confirm the wave-like nature of light.

When a beam of electrons is fired towards a similar pair of slits, classical physics predicts only particles on certain paths will pass through the slits ...

Electron gun

... but instead the electrons create an interference pattern. This phenomenon shows that electrons, too, exhibit wavelike properties.

De Broglie's hypothesis

In 1924, French physicist Louis de Broglie put forward a radical hypothesis that was to lay the foundation for later developments in quantum physics. Up until this time, the wave-particle duality of light had been considered as just another odd feature of electromagnetism. De Broglie, however, suggested that duality might work in the opposite direction too – particles such as the electron might be carried by an associated wave. De Broglie showed that the 'wavelength' λ of a particle with momentum p could be calculated using the equation:

$$\lambda = \frac{h}{p} \quad , \quad \text{or:} \quad \lambda = \frac{h}{mv}\sqrt{\frac{1-v^2}{c^2}}$$

where m is the particle's 'rest mass' (see page 374), v its velocity, c is the speed of light and h is Planck's constant.

De Broglie's daring hypothesis was soon confirmed by the discovery of electron diffraction (see page 292), although it took some time for its full implications to become clear.

$$\lambda = \frac{h}{mv}$$

Electron microscopy

As its name suggests, an electron microscope is a device that uses electrons, rather than light, to study very small objects. It relies entirely on the fact that electrons can display quantum wave-like properties with wavelengths around 100,000 times smaller than those of visible light.

Transmission electron microscopes (TEMs), first used in the early 1930s, fire a beam of energized electrons (a cathode ray – see page 254) through a thin sample of material. Here, they are deflected by wave phenomena such as diffraction and scattering, before producing an image on a screen or photographic film. TEMs can achieve magnifications of up to 10 million times. Scanning electron microscopes, developed in the 1950s, pass a narrow beam back and forth across the surface of a specimen many times, while detectors measure changes in the scattered light. Their magnifications are limited to a maximum of around 1 million times, but their ability to image relatively large samples makes them very useful.

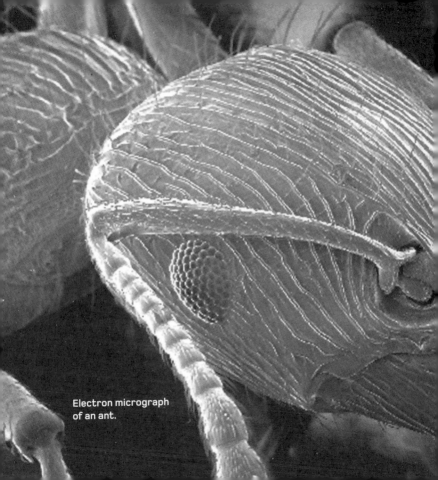

Electron micrograph
of an ant.

Schrödinger wave equation

For some years after de Broglie suggested particles could exhibit wave-like properties (see page 290), debate raged about what this actually *meant*. German physicist Max Born soon demonstrated that the shape of the 'matter wave' was linked to the likelihood of a particle being observed in a given point in space – the stronger the wave at that point, the more likely the particle was to be found there. Austrian scientist Erwin Schrödinger's wave equation, published in 1926, described the particle's waveform (its distribution in space) for the first time.

However, even with the waveform established, doubts remained about whether it was purely a predictive tool for measuring the behaviour of classical, point-like objects (as suggested by the Copenhagen interpretation – see page 298), or whether it had a deeper signficance. Schrödinger argued for the objective reality of the wavefunction – in other words, the properties of particles really *are* 'smeared out' across infinitesimal volumes of space, just as his equations describe.

$$H(t) \mid \psi(t) \rangle = i\hbar \frac{d}{dt} \mid \psi(t) \rangle$$

Quantum tunnelling

One of the earliest and most successful applications of de Broglie's hypothesis was the discovery of quantum tunnelling – the ability of wave-like quantum particles to pass through barriers that are insurmountable in classical physics.

Tunnelling was discovered through studies of alpha decay (see page 260) in particular. In effect, for an alpha particle (helium nucleus) to escape from a heavier atomic nucleus, it must overcome the energy binding it to the rest of the nucleus. But by the 1920s it was clear that a classical description of the particle could never produce enough energy to escape the 'potential well' created by the binding energy. In 1928, researchers showed that the Schrödinger equation could be applied to an alpha particle to allow it to 'tunnel' out of the nucleus, explaining both the frequency of the emission process (see page 268), and the energies of escaping particles. Tunnelling has since proved to have other applications – it plays a key role in the design of modern semiconductors and transistors.

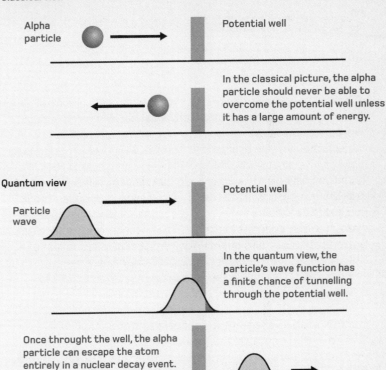

Classical view

Alpha particle → Potential well

In the classical picture, the alpha particle should never be able to overcome the potential well unless it has a large amount of energy.

Quantum view

Particle wave → Potential well

In the quantum view, the particle's wave function has a finite chance of tunnelling through the potential well.

Once throught the well, the alpha particle can escape the atom entirely in a nuclear decay event.

Copenhagen interpretation

Developed at the University of Copenhagen in the mid-1920s by Danish physicist Niels Bohr and his German colleague Werner Heisenberg, the Copenhagen interpretation is the most famous of several models for the way in which quantum physics interacts with the large-scale universe. It treats the quantum wavefunction (see page 294) as a purely mathematical tool describing the likelihood of observing a particular result – for instance, finding a particle in a certain location. The act of 'observation' (broadly encompassing any interaction between the macroscopic and quantum worlds) triggers a 'collapse of the wavefunction' – settling on a particular outcome out of the many possibilities, so that only a single result is observed.

This interpretation predicts important aspects of quantum physics such as the uncertainty principle (see page 300), but Erwin Schrödinger expressed his doubts about it (see page 302), and many modern physicists prefer alternatives, such as the 'many worlds' interpretation (see page 304).

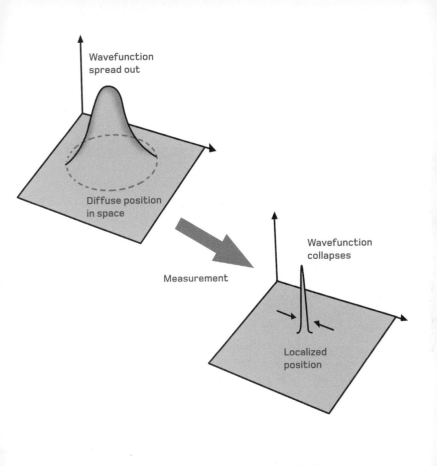

Wavefunction
spread out

Diffuse position
in space

Measurement

Wavefunction
collapses

Localized
position

Quantum mechanics and the uncertainty principle

In the mid-1920s, German physicist Werner Heisenberg developed an approach to quantum physics called matrix mechanics (or quantum mechanics), in which the motion of particles was described using a mathematical technique called a Fourier transform. One result of this approach was that certain pairs of particle properties could not be determined simultaneously with absolute precision. Heisenberg later formalized this 'uncertainty principle' in the equation:

$$\Delta x \Delta p \geq h$$

where Δx is the uncertainty in measurement in position, Δp is uncertainty in momentum and h is Planck's constant. According to this equation, the more exact the measurement of position, the greater the error in momentum and vice versa. Heisenberg illustrated the principle with a thought experiment, described opposite, blaming uncertainty on the effects of observation and measurement. However, today it is widely accepted as an intrinsic property of the quantum world.

1 Uncertainty of position: the more accurately the object's wavelength is known the less accurately its position can be determined

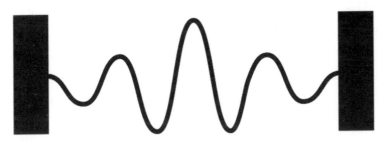

2 Uncertainty of wavelength: the more tightly constrained the object's location, the harder it is to pinpoint its wavelength

Schrödinger's cat

Perhaps the most famous thought experiment in all of physics was outlined by Erwin Schrödinger in 1935 in order to highlight his concerns about the Copenhagen interpretation of quantum theory (see page 298). Illustrated opposite, the experiment attempts to map the quantum-level uncertainty of a radioactive decay event (see page 259) into a macroscopic situation – namely whether a cat in a box is alive or dead.

Schrödinger argued that, if Heisenberg's observer-driven ideas were correct, the cat would exist in a superposition of states, alive and dead at the same time. However, this is now recognized as an extreme version of the Copenhagen interpretation – Niels Bohr was already arguing that the wavefunction of such a system would collapse the moment the detector measured a decay event with no need for an external observer (so-called 'objective collapse'), and so there would never be a macroscopic superposition. Other interpretations resolve the problem in different ways (see page 304).

Schrödinger imagined an experiment in which a cat is locked in a box with a radioactive source that has a 50/50 chance of decaying in a certain time. If the source decays, then it triggers the release of a deadly poison, but if the nucleus does not decay, the cat lives. The superposition of quantum states in the source suggested by the Copenhagen interpretation implies that until the system is observed, the cat is both alive and dead.

Other quantum interpretations

Although the Cophenhagen interpretation of quantum theory is probably the best known, physicists have developed several other important approaches to the problem of how quantum uncertainty affects the real world we perceive on a macroscopic scale. Often these rely on the idea of 'decoherence' – the idea that the wavefunction of a quantum system can slip out of phase and give the appearance of collapse when it is measured, without actually collapsing.

In the late 1950s, US physicist Hugh Everett III used the idea of decoherence to develop his 'many worlds' interpretation. Beloved of science fiction authors, this suggests that every possible state of the wavefunction interacts with its environment, and through quantum entanglement (see page 316) produces countless mutually unobservable alternate Universes – one for each possible state of each possible quantum event. Within any given Universe, the wavefunction appears to collapse in a certain way because only one state can be seen.

Other interpretations are perhaps a little less radical. For instance, the 'consistent histories' approach uses some complex mathematical ideas to 'generalize' the Copenhagen interpretation, avoiding wavefunction collapse but instead allowing different probabilities to be assigned to the possible 'histories' of an entire system – on both quantum and classical scales. Unlike the many-worlds interpretation, however, the consistent histories approach does not suggest that all the different possible histories occur, or allow predictions of which histories will occur – it merely describes the Universe as we observe it.

Finally, the ensemble or statistical interpretation has its own unique take on the meaning of the wavefunction: rather than viewing it as a feature of an individual quantum system, it treats it instead as a description of probabilities across a hypothetical huge array or ensemble of identical systems. Each particular system within the ensemble has only one possible state, and it is the positon of the system within the ensemble that the wavefunction describes. Although somewhat out of favour at present, this was Albert Einstein's preferred understanding of the way in which quantum physics affects the 'real' Universe.

Quantum numbers

One of the many strange aspects of quantum theory is the fact that many properties turn out to be 'quantized' on a subatomic scale – rather than varying continuously in the way we might expect from the classical physics, they take on discrete values in integer or half-integer units, with no intermediate values between them. Properties that behave in this way are said to display 'quantum numbers'.

One of the most important sets of quantum numbers are those describing the placement of electrons in orbital shells around the atomic nucleus. The principal quantum number n is the number of the shell from the inside out, while azimuthal, magnetic and 'spin projection' quantum numbers l, m_l and m_s all help define an electron's placement within that shell. The values of each number are limited as shown opposite. Rules governing relations between elementary particles are often related to their own quantum numbers, perhaps the most significant of which is the property called 'spin' (see page 308).

n	Possible values of l	Subshell designation	Possible values of m_l	Number of orbitals in subshell
1	0	1s	0	1
2	0	2s	0	1
	1	2p	1, 0, -1	3
3	0	3s	0	1
	1	3p	1, 0, -1	3
	2	3d	2, 1, 0, -1, -2	5
4	0	4s	0	1
	1	4p	1, 0, -1	3
	2	4d	2, 1, 0, -1, -2	5
	3	4f	3, 2, 1, 0, -1, -2, -3	7

n = principle quantum number
l = azimuthal quantum number
m_l = magnetic quantum number

NB – two electrons displaying opposite spin numbers m_s can occupy the same orbital

Spin

Spin is a unique form of angular momentum (see page 34) displayed by particles susceptible to the laws of quantum physics, including atoms and subatomic particles. It is analogous (but not identical) to the rotational angular momentum displayed by a classical rotating object spinning on its axis, such as a spinning top or a planet. It manifests as an additional form of angular momentum that can be identified separately from the orbital angular momentum displayed by particles within atoms, and has its own 'spin quantum number' (s), that is quantized in either whole-integer or half-integer steps.

Spin is significant because it affects a broad variety of particle behaviours, and divides elementary particles into two major groups known as fermions and bosons (see page 310). In association with electric charge, it is also responsible for creating the magnetic dipole moments (see page 194) of individual particles.

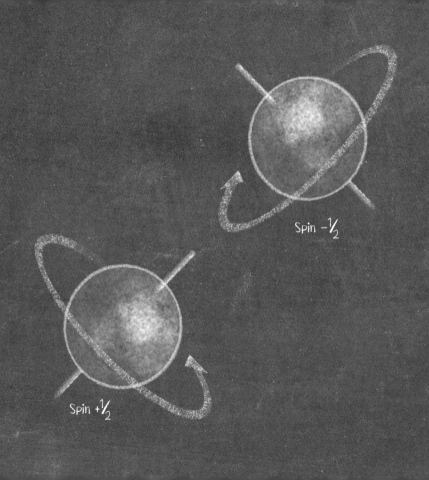

Spin −½

Spin +½

Bosons, fermions and the Pauli exclusion principle

The property of spin (see page 308) divides both elementary particles and atoms into two distinct families. Bosons are those with zero or integer (whole-number) spin, while fermions are those with half-integer spins, such as $\frac{1}{2}$, $\frac{3}{2}$ and $\frac{5}{2}$. All the elementary particles of matter (hadrons and leptons, see pages 342 and 330) are fermions, but when combined together in larger particles or atoms, they can produce bosons.

An important distinction between the groups is that fermions obey the Pauli exclusion principle. Discovered by Austrian physicist Wolfgang Pauli in 1925, this rule states that no two identical fermions in a system can occupy the same quantum state (described by their quantum numbers) simultaneously, explaining, for example, why the electrons in an atom all occupy separate orbital positions and energy levels. Bosons, meanwhile, obey a set of rules called Bose-Einstein statistics, which allow many particles to exist in identical states (as seen, for example, in many superfluids – see page 320).

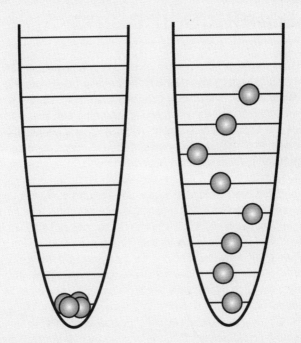

Bosons in a system can all 'fall' into the same energy state.

Fermions are forced to occupy unique energy levels.

Magnetic resonance imaging

By making use of the quantum property known as spin (see page 308), magnetic resonance imaging (MRI) scanners allow scientists and doctors to study the soft tissues of the human body in three dimensions, overcoming the limitations and risks of traditional X-ray photography. The MRI technique makes use of the fact that our bodies are dominated by water molecules, and so are rich in hydrogen nuclei, most consisting of just a single proton. When placed in a strong magnetic field, the individual magnetic moments of these protons line up, and a pulse of radio waves can generate an electromagnetic field with the precise energy (or 'resonant frequency') required to invert or 'flip' their spin. Once the field is removed, the nuclei revert to their original states, releasing radio-frequency signals that can be detected and mapped. A wide range of different factors determine the rate at which the nuclei flip back, allowing doctors to identify specific tissues with certain characteristics. 'Contrast agents' can also be used to intensify the behaviour of certain tissues and make them stand out.

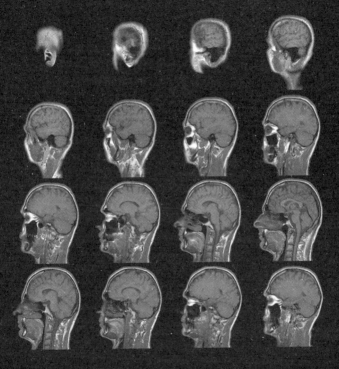

A series of MRI scans highlights features of a patient's head.

Degeneracy

In normal circumstances, the position of a subatomic particle within an atom determines its 'energy level'. When two particles occupy the same energy level in the same system, they are said to be degenerate – they can only exist in the same energy level if they have some other properties (with their own quantum numbers) to distinguish them from each other. For example, all the electrons in a particular orbital shell (with a quantum number n – see page 306) have essentially the same energy level, but three additional quantum numbers – l, m_l and m_s – allow multiple electrons to exist in degenerate states.

In superdense conditions (for instance, in the collapsed stellar cores known as white dwarfs and neutron stars), quantum effects generate a pressure that repels particles from each other. White dwarfs (with a solar mass of material compressed into a volume the size of Earth), are saved from total collapse by 'degenerate electron pressure', while the even denser neutron stars are held up by degenerate neutron pressure.

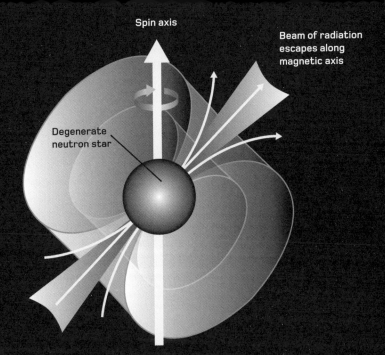

Spin axis

Beam of radiation
escapes along
magnetic axis

Degenerate
neutron star

A pulsar or pulsating star is a degenerate neutron
star – a collapsed stellar core with a mass of more
than 1.5 suns – that beams radiation across space
like a cosmic lighthouse.

Entanglement

Perhaps the strangest of all the many weird concepts in quantum physics, entanglement is a phenomenon in which the quantum properties of two particles become co-dependent, so that the properties of one are instantaneously affected by measurements conducted on the other.

Entangled systems require special preparation – for example, it is possible to physically create a pair of electrons that must have opposite spins (in accordance with the Pauli exclusion principle – see page 310), while the actual spin of each particle remains in a state of quantum uncertainty (a situation Erwin Schrödinger described as 'entanglement of the wavefunction'). When the pair are separated (even by a huge distance) and one particle's spin is measured, the spin of the other particle will automatically resolve itself in the other direction. The effect happens instantaneously, apparently breaching the speed of light and the rules of relativity – little wonder Einstein referred to it as 'spooky action at a distance'.

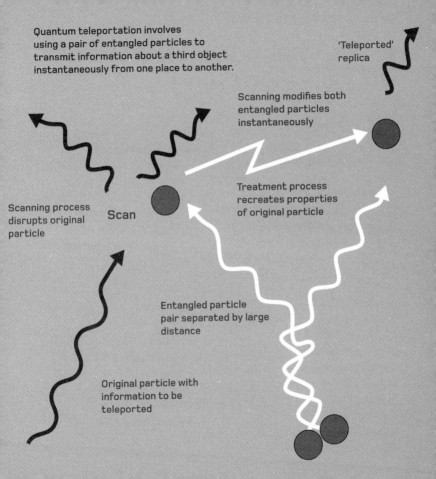

Quantum teleportation involves using a pair of entangled particles to transmit information about a third object instantaneously from one place to another.

'Teleported' replica

Scanning modifies both entangled particles instantaneously

Scanning process disrupts original particle

Scan

Treatment process recreates properties of original particle

Entangled particle pair separated by large distance

Original particle with information to be teleported

Quantum computing and cryptography

The field of quantum computing uses quantum effects to manipulate and process information. It has tremendous potential, since while a binary 'bit' of classical data can take only one value at a time, a quantum 'qbit' can represent any and all possible values in a given range: until it is measured, it exists in a 'superposition' of all possible states. Quantum computers are particularly well suited for solving certain problems that can at present only be solved by 'brute force' processor power – an array of a few dozen qbits, for example, can store more than a terabyte of conventional data.

A closely related field is quantum cryptography, which uses quantum phenomena to send secure messages between recipients. By encoding information using quantum techniques, it is possible to create a situation in which anyone attempting to read (measure) the information will disturb it in a way that can be detected – this is the secret behind 'quantum key distribution', the most successul approach so far developed.

One way physicists model the exact mixing of 1 and 0 in a quantum 'qbit' is to think of it as the latitude on a sphere, with a 'north pole' (1) equivalent to a value of 1 and a 'south pole' (2) equivalent to value 0. Superpositions – a mix of 1 and 0 values – can be considered as intermediate latitudes (3). The measurement process collapses the qbit into a classical value of 1 or 0, with the probability of each value given by the surface area on the opposite side of qbit's latitude (4).

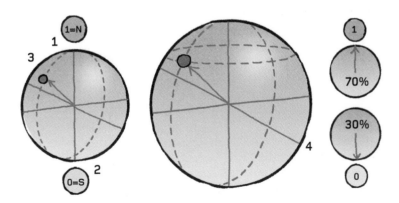

Superfluids

When cooled to temperatures close to absolute zero, materials such as liquefied gases can demonstrate a strange behaviour in which they lose all their internal viscosity or friction. Known as superfluidity, this unusual state of matter is somewhat analogous to superconductivity (see page 250). The phenomenon arises when all the particles within the material fall into a single shared quantum state and can therefore act as if they are a single enormous, well-ordered particle. The first superfluid to be identified was liquid helium, discovered by physicists at the University of Cambridge in 1937.

The majority of superfluids are Bose-Einstein condensates, formed from gases in which the atoms display an overall spin of either zero or an integer value (in other words, they are bosons – see page 310). However, fermionic particles (with half-integer spins) can also become superfluids in a number of ways: the most common method involves coupling together as 'Cooper pairs' whose overall spins have integer or zero value.

A computer simulation models the gradual development of a Bose–Einstein condensate superfluid as large numbers of atoms begin to acquire identical quantum properties and start to behave like a single particle.

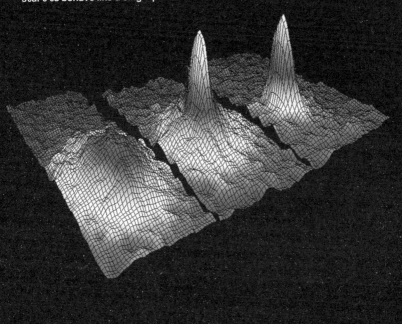

Particle physics

The quest to discover the fundamental constitutents of matter, the rules that govern their behaviour and the forces that bind them together and allow them to interact is one of the driving projects of modern physics, pushing at the frontiers of both complex mathematics and technology. Fortunately, while the research behind particle physics is exhaustive and the reasoning behind its conclusions often daunting, many of its actual findings (and even some of its predictions) are surprisingly intuitive.

At its heart, particle physics is concerned with interactions between indivisible elementary particles that are mediated by fundamental forces. Of the three most common subatomic particles – the proton, neutron and electron – only one, the electron, is truly elementary (see page 162). Particle accelerators (see page 326) can be used to break the heavier constituents apart, releasing showers of hidden particles that often display a confusing mix of properties. However, it eventually becomes clear that many of these complexities are caused by a limited number of particles combining in a multitude of different ways. Once all the complexities are cleared away, there is plentiful evidence to support an elegant pattern known as the Standard Model (see page 324).

The other major challenge for particle physics lies in uniting the four fundamental forces that govern interactions between particles, and this has proved to be rather more problematic. Three of these forces – the electromagnetic, weak and strong nuclear forces (see pages 334, 346 and 342) – share common features that allow them to be described by 'quantum field theories', and there are plausible hopes that they can be explained as aspects of a single 'unified field theory', but the fourth – gravity – appears to be radically different and prospects for a 'Theory of Everything' are rather less advanced. Nevertheless, the complex ideas of string theory, with its attendant higher dimensions (see pages 358 and 360) offer a tantalizing prospect of a theory of everything that would not only unify all the fundamental forces, but also explain the varied properties of elementary particles.

The Standard Model

The most successful modern approach to particle physics was pioneered by US physicists Sheldon Glashow, Steven Weinberg and Abdus Salam in the 1960s and took on its present form in the 1970s after several of its initial predictions had been confirmed by particle accelerators of the time (see page 326). Now known as the Standard Model, it predicts a surprisingly simple pattern of elementary particles.

According to the model, matter consists of three 'generations' of fermions (particles with half-integer spins), at different levels of energy and mass. Each generation contains a pair of quarks and a pair of leptons (see pages 328 and 330), and each of the matter particles has an antimatter equivalent (omitted from the chart opposite for clarity). Forces, meanwhile, are transmitted by five additional 'gauge bosons' (see page 338), including the positively and negatively charged W bosons. A final particle, the Higgs boson, is associated with the Higgs field that gives matter its mass (see page 348).

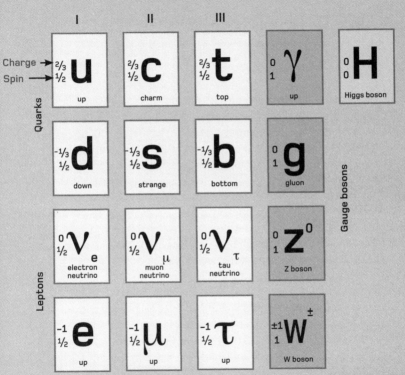

Fermion generations

	I	II	III		

Charge →
Spin →

Quarks

2/3 **u** 1/2
up

2/3 **c** 1/2
charm

2/3 **t** 1/2
top

0 **γ** 1
up

0 **H** 0
Higgs boson

-1/3 **d** 1/2
down

-1/3 **s** 1/2
strange

-1/3 **b** 1/2
bottom

0 **g** 1
gluon

Leptons

0 **ν**e 1/2
electron neutrino

0 **ν**μ 1/2
muon neutrino

0 **ν**τ 1/2
tau neutrino

0 **Z**⁰ 1
Z boson

Gauge bosons

-1 **e** 1/2
up

-1 **μ** 1/2
up

-1 **τ** 1/2
up

±1 **W**± 1
W boson

Particle accelerator

As its name suggests, a particle acclerator is a machine designed to accelerate atomic and subatomic particles to high speeds (close to the speed of light) before colliding them with each other. Despite their nickname of 'atom smashers', these machines do not actually split particles into their smaller subcomponents. Instead, the violence of the impact effectively destroys the particles, releasing a burst of energy that is swiftly converted back into new particles in accordance with Einstein's famous equation $E=mc^2$. A variety of detector instruments track the paths of these particles in order to reveal properties such as their mass, electric charge and spin.

Accelerators typically use electromagnetism to transfer energy into particles – small-scale acceleration may simply involve firing particles into a high-voltage electrostatic field, but high-energy accerators typically use powerful electromagnets arranged in a linear or circular configuration, as seen at the Large Hadron Collider (see page 362).

The Large Hadron Collider

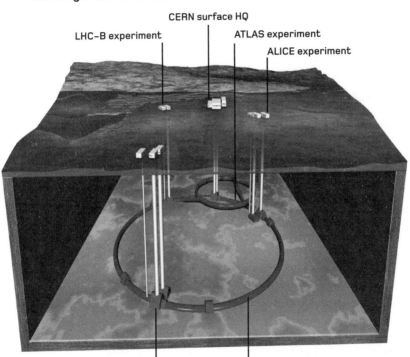

CERN surface HQ

LHC-B experiment

ATLAS experiment

ALICE experiment

Compact Muon Solenoid (CMS) experiment

Main ring 27 km, up to 175 m below surface

Quarks

Quarks are the fundamental constituents of heavy subatomic particles such as protons and neutrons. They are never seen in isolation, but their theoretical existence was independently proposed in 1964 by US physicists Murray Gell-Mann and George Zweig, and subsequently proved by particle accelerator experiments.

Experiments have shown that there are six different types or 'flavours' of quark, known as up, down, strange, charm, top and bottom quarks. Of these, the up and down quarks are by far the most stable, and it is these two that make up protons and neutrons (as shown opposite). In general, quarks bind together either in triplets to form 'baryons' or in quark-antiquark pairs to form 'mesons'. All quarks have a spin of ½ and are thus fermions, affected by the Pauli exclusion principle (see page 310). They have electric charges of either +⅔ or −⅓, and also possess a 'colour charge' that allows them to interact via the strong nuclear force (see page 342).

Proton

Neutron

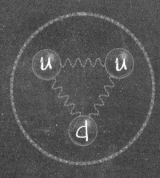

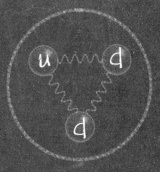

2 x up quark
1 x down quark
charge = 1
spin = ½

1 x up quark
2 x down quark
charge = 0
spin = ½

Leptons

Matter particles that are not influenced by the strong nuclear force are known as leptons. The best known is the electron, but experiments have revealed that there are in fact three 'generations' of leptons, each consisting of a negatively charged particle and an associated neutrino (see page 266). The electron is the lightest and most stable of the charged leptons and is associated with the 'electron neutrino' (the only form of neutrino that can be directly detected). A second generation is formed by the heavier, unstable muon (μ^-) and its muon neutrino, while the third-generation particles are the tau (τ^-) and tau neutrino. Leptons have a spin of ½ and are therefore fermions – they lack 'colour charge' and are immune to the strong nuclear force, but are still influenced by the electromagnetic and weak nuclear forces, as well as gravity. As with all forms of matter, leptons have equivalent antiparticles, with either reversed (positive) charge in the case of the charged leptons, or reversed 'chirality' (a property related to their spin) in the case of uncharged neutrinos.

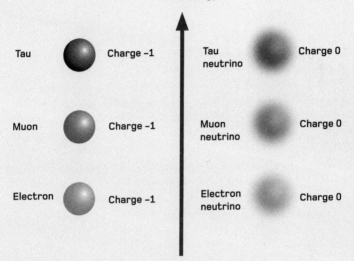

Increasing
mass/energy

Tau   Charge –1   Tau   Charge 0
neutrino

Muon   Charge –1   Muon   Charge 0
neutrino

Electron   Charge –1   Electron   Charge 0
neutrino

Fundamental forces

According to the Standard Model, four forces are responsible for all the interactions that bind matter together. Perhaps the most familiar of these is the electromagnetic force, which acts on particles with electric charges or magnetic moments. Although relatively weak, electromagnetism can act over large distances, and hence reveals its influence in many aspects of the world around us.

The strong and weak nuclear forces, in contrast, both operate on such small scales that their effects are confined within the atomic nucleus. The weak force affects all fermions, and has a number of unusual features compared to the other forces, while the strong force affects only hadrons (quarks and particles containing them).

Finally, the gravitational force is a strange outlier among the forces – it is extremely weak and only makes its influence felt where large accumulations of matter are involved, but it does operate over all distances and is therefore crucial to shaping the structure of the large-scale Universe.

By studying how particles behave in particle accelerators (see page 326) it has become clear that the electromagnetic, strong and weak forces grow increasingly similar at higher energies. Above a certain threshold, electromagnetism and the weak force can be described by a single 'electroweak' interaction, and theoretical physicists hope that the strong and electroweak forces can be unified for even higher energies to create a Grand Unified Theory (see page 350). Gravitation, however, remains a troublesome outlier – the predictions of Einstein's theory of general relativity, which describes it so successfully in other ways, are often at odds with those of quantum physics.

Despite these problems, most physicists believe that the four forces will ultimately prove to be reconcilable in a single 'superforce' that existed for a brief instant in the furnace-like conditions of the Big Bang that created the Universe (see page 398). The forces then rapidly fragmented, with gravitation splitting away first to explain its radical differences, to produce the four distinct aspects we see today.

Electromagnetic force

The electromagnetic force is responsible for many of the familiar phenomena we observe in the Universe around us. It manifests itself as light and other radiations, and in electrical and magnetic fields. It is also capable of influencing susceptible objects (those with electric charges or magnetic moments) over long ranges.

In classical physics, we usually consider electromagnetism as a field effect – certain objects generate electromagnetic force fields around them, with 'lines of force' that influence objects within the field. But how exactly does a particle 'know' that it should react to an electromagnetic force? This is a thorny question because electromagnetism has so many different aspects – and since light is an electromagnetic wave, any theory of electromagnetism also needs to be consistent with the theory of special relativity that describes light's behaviour. The most successful explanation for how electromagnetic force is transmitted between particles is known as QED (see page 336).

Quantum electrodynamics

Developed by Japanese physicist Sin-Itoro Tomonaga and Americans Julian Schwinger and Richard Feynman in the 1940s and 1950s, quantum electrodynamics, or QED, is a 'relativistic field theory' describing the electromagnetic force. It was the first theory to successfully unite special relativity with quantum physics, and paved the way for similar 'gauge theories' attempting to explain other forces (see page 344).

At its heart, QED describes how electromagnetic force is transferred between objects by the exchange of photons. In this situation, the photons act as 'gauge bosons' and need not be detectable in any measurable sense (see page 338). Although mathematically complex, the fundamentals of QED are made relatively easy to grasp thanks to the diagrammatic approach to the subject developed by Richard Feynman. In essence, this involves considering all the possible quantum interactions between photons and charged particles in a system, and then mapping out and summing the probabilities of these events.

This simple Feynman diagram shows
how interactions between particles
(in this case electrons) are mediated
by the exchange of a gauge boson, in
the form of a virtual photon.

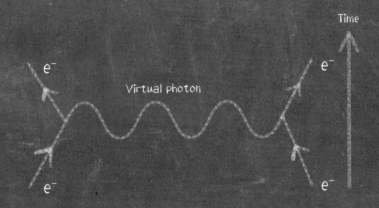

Time

e^-

e^-

Virtual photon

e^-

e^-

Gauge bosons

In particle physics, a gauge theory is one in which forces are transmitted between matter particles (composed of individual fermions with half-integer spins) by means of force-carrying 'gauge bosons' (particles with zero or integer spin). These include photons (carrying the electromagnetic force), W and Z particles (carrying the weak nuclear force) and gluons (transmitting the strong nuclear force between quarks).

During fundamental interactions, gauge bosons are exchanged in both directions, but importantly this may be done via 'virtual' particles. These are bosons that can pop into existence for very brief periods thanks to the 'time-energy uncertainty relation' – an aspect of the uncertainty principle (see page 300) that permits the instantaneous creation of particles with limited energies (or masses) for a very limited time. Extreme situations (for example, in particle accelerators and around black holes) can allow heavier particles to persist for longer periods of time, and even to become 'real' and detectable.

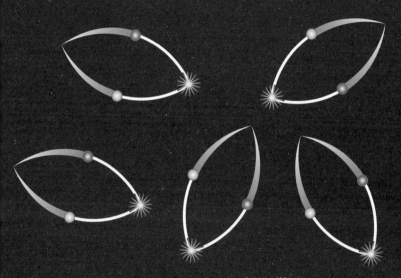

According to the time-energy uncertainty relation, it's possible to 'borrow' an amount of energy ΔE from the vacuum of space for a very short period of time Δt.

$$\Delta E \times \Delta t < \hbar/2$$

where ℏ is the 'reduced Planck constant'. These create a virtual particle-antiparticle pair that can act as gauge bosons between interacting particles of 'real' matter.

The Casimir effect and zero-point energy

Some of the best evidence for the existence of the 'virtual particles' associated with QED and other quantum field theories (see pages 336 and 338) comes from the Casimir effect. Discovered by Dutch physicist Hendrick Casimir in 1948, this is a phenomenon driven by the spontaneous creation of virtual photons, which produces a minute attractive force between two narrowly separated, uncharged metal plates in a vacuum. It can also be used to generate a repulsive force.

The effect's existence and strength have been confirmed in several experiments, and it is now viewed as a background energy inherent to quantum field theories. Even at the lowest energy state of a quantum system, it seems, there is still some 'zero-point' or 'vacuum' energy. Zero-point energy may be the origin of the mysterious 'dark energy' driving the expansion of the Universe (see page 404). It could offer a potentially huge source of energy, and might offer ways of creating 'exotic' matter to keep wormholes open (see page 388).

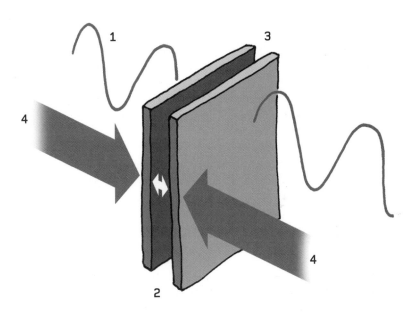

1 Long wavelengths of light fill the space around plates

2 Plates are separated by a fraction of the wavelength of light

3 Light cannot pass between closely separated plates

4 Pressure from the energy density difference pushes plates together from outside

Strong nuclear force

The strong nuclear force is the most powerful of all the fundamental forces, but only operates over extremely short ranges and is only felt between particles within the confines of the atomic nucleus. The force operates only between hadrons (particles that possess a property known as colour charge), and is thus limited to the elementary quarks (see page 328) and mesons or baryons – particles made up of two or three quarks respectively.

Strong force interactions can take two distinct forms, and some physicists insist that the term 'strong force' should be strictly limited to the extremely short-range force binding directly between quarks *within* mesons and baryons. In this environment, the force is carried by gauge boson particles known as gluons. On slightly larger scales, the so-called 'residual strong force' operates between mesons and baryons and is carried by particles called pions, which are themselves a form of meson containing an up and an anti-down quark.

The strong force binds protons (1) and neutrons (2) in nuclei, such as those shown here, as well as the 'up' (3) and 'down' (4) quarks within individual protons and neutrons.

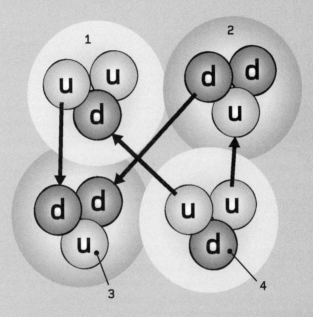

Quantum chromodynamics

The quantum field theory of the strong nuclear force is known as quantum chromodynamics (QCD). Like quantum electrodynamics (see page 336), it is a gauge theory in which force is transmitted between susceptible particles by messenger particles (see page 338). The force-carriers between individual quarks are gauge bosons known as gluons, while those acting between larger particles are 'pi mesons' or pions.

In QCD, strong-force interactions are governed by a property called colour charge. The name is somewhat misleading, since the property has nothing to do with visual colour, and only superficially resembles electric charge. Quarks can take one of three distinct colours – red, green or blue (antiquarks have equal and opposite colours – 'anti-red' and so on). The three quarks within a baryon must have different colours that cancel out to make it appear 'white' from the outside. However, the cancellation is not perfect and some colour 'leaks out' to make baryons susceptible to the residual strong force.

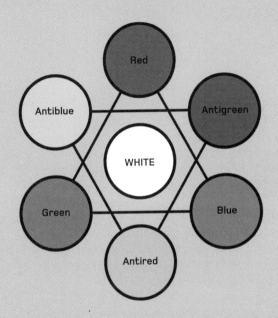

In quantum chromodynamics, quarks form 'colour charge' combinations that appear 'white' from outside – either baryons containing quarks of all three colours or 'anticolours', or quark–antiquark pairs (mesons) containing a single colour and its anticolour.

Weak nuclear force

As its name suggests, the weak nuclear force acts on the limited scale of the atomic nucleus and is significantly weaker than the strong force. However, it has several other important and unique features: it affects all fermions (elementary 'matter' particles – see page 324), it is the only interaction capable of changing the type or 'flavour' of quarks (see page 328), and it is 'asymmetric' (see page 353).

As with the electromagnetic and strong nuclear forces, the weak force is best described by a gauge theory, in which it influences fermions according to an innate property known as weak isospin (denoted T_3) and is transmitted through space by gauge bosons (see page 338) called W^+, W^- and Z.

These bosons are unusual in several ways. Most obviously, there are three of them (electromagnetism and the strong force require one each). Furthermore, the W^+ and W^- particles are electrically charged. This gives rise to two different types of weak-force influence, known as neutral-current and charged-current interactions. Neutral-current interactions resemble other forces – they involve simple exchange of neutral Z bosons and transmission of force – while charged-current interactions also involve a transformation of affected particles from one type to another.

For instance, a charged lepton such as an electron can absorb a W boson of opposite charge to transform into its corresponding neutrino:

$$e^- + W^+ \longrightarrow \nu_e$$

Quarks can also change their electric charge and flavour by shedding or absorbing W bosons. Most commonly, a down quark (d) can change into an up quark (u) by releasing a W⁻ particle:

$$d \longrightarrow u + W^-$$

This is the mechanism responsible for the form of radioactivity known as beta decay in which a neutron spontaneously changes into a proton (see page 262). In that situation, the W⁻ particle then decays rapidly to produce an electron and an electron antineutrino, the typical products of beta decay.

The Higgs field and Higgs boson

The mass of the W and Z bosons that transmit weak nuclear force presents a huge problem for physicists, since gauge bosons are supposed to be massless. In 1964, British physicist Peter Higgs and others proposed the existence of an unusual field permeating space, which could interact with the bosons via a 'Higgs mechanism', breaking certain symmetries (see page 352) and giving rise to the mass of the W and Z bosons. Further research showed that fermions (matter particles) could also gain their mass through interactions with the Higgs field and predicted the existence of an associated particle, the Higgs boson, with no electric charge, no spin and no colour charge. Both field and boson were soon adopted as cornerstones of the Standard Model: identifying the Higgs boson would prove the existence of the field and show that the Standard Model is correct, but detecting this massive, short-lived and highly unstable particle proved an enormous challenge. In 2013, tentative evidence for its existence emerged from experiments at the Large Hadron Collider (see page 362).

Simulation of a Higgs particle 'event' in the Large Hadron Collider.

Grand Unified Theory

A Grand Unified Theory (GUT) is the catch-all term in particle physics for any theory describing a single overarching 'electronuclear force' that attempts to unify electromagnetism with the weak and strong nuclear forces. The relationship between these forces would only become apparent at extremely high energies, where the electronuclear force would govern all subatomic interactions (since gravity is so weak on these scales compared to the other forces).

Convincing support for a GUT comes from the fact that the weak nuclear force and electromagnetism have already been unified in an 'electroweak' theory (see page 333). Moreover, the nuclear forces and electromagnetic force can all be viewed as belonging to a complicated mathematical 'object' called a Lie group, which exhibits certain symmetries known as gauge symmetries (hence the term 'gauge boson'). This similarity between the forces is one of the most promising pieces of evidence for the idea that they can be unified.

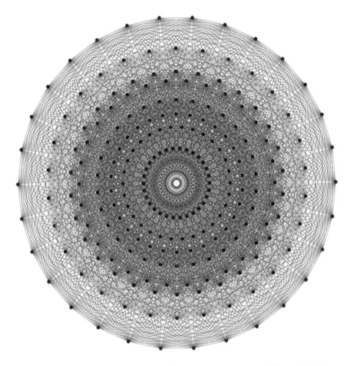

Computer model of the so-called 'E8' Lie group – a mathematical 'object' based on studies of symmetry that is believed to describe both electromagnetism and the weak and strong nuclear forces.

Symmetry

One of the most important concepts in particle physics is an extension of the geometrical concept of symmetry. A physical object is said to be symmetrical when it remains essentially unchanged after an operation called a transformation (typically rotating or flipping its coordinates), while a physical law or an interaction is symmetrical if it remains unchanged following a change to the characteristics of the particles involved.

The Standard Model (see page 324) involves three distinct symmetries: charge symmetry (C-symmetry), in which each particle is replaced with its antiparticle; parity symmetry (P-symmetry), in which the orientations and spins of particles are reversed to create mirror-image particles; and time symmetry (T-symmetry), in which the flow of time is reversed. Certain processes are symmetrical when subjected to combinations of two such transformations (for example, charge-parity or CP-symmetry), and the Standard Model insists that a combination of all three transformations is symmetrical in all circumstances (CPT symmetry).

When a process does not exhibit symmetry, it is said to be in violation of symmetry for a particular transformation. For example, weak nuclear force interactions involving the W^+ and W^- bosons (see page 346) break both parity symmetry and CP-symmetry. CP violation is a particularly rich area of research for modern physics because a CP-breaking process early in the history of the Universe is the most likely cause of the present-day imbalance between matter and antimatter (see page 168).

Symmetry is also a key area of study in the quest for a Grand Unified Theory (see page 350), and some theoretical physicists believe that a key stage on the route to this would be the discovery of 'supersymmetry'. This is an additional type of symmetry in which each fermion from the Standard Model would have a boson counterpart called a 'superpartner', and each Standard Model boson would likewise have a symmetrical fermion superpartner. Despite the elegance of the theory, however, there is so far no firm evidence that these superpartner particles actually exist.

Graviton

While electromagnetism and the weak and strong nuclear forces can all be described using quantum field theories, gravity remains an outsider – its behaviour is best described by the general theory of relativity (see page 380), in which it is seen as a disturbance to the fabric of spacetime itself. Efforts to explain gravity on the same terms as the other forces give rise to a model of 'quantum gravity' in which the force is transmitted by a hypothetical gauge boson called the graviton.

Although the graviton has not been detected and may lie well beyond the reach of the most powerful particle accelerators, physicists can make surprisingly precise predictions about its properties. It must be a massless particle with a spin of 2 (compared to the photon, whose spin is 1), travelling at the speed of light. In classical situations, the graviton's behaviour should 'reduce' to match the predictions of general relativity, allowing it to manifest phenomena such as distortion of spacetime and gravitational waves (see page 384).

Theories of Everything

Any attempt to describe all four fundamental forces as aspects of a single united force is known as a Theory of Everything (TOE). Any TOE must reconcile the quantum field theories that describe electromagnetic and nuclear forces with the behaviour of gravity, described by general relativity as a distortion in the fabric of spacetime itself (see page 378). Another major challenge is the mathematical problem of handling 'infinities' that arise naturally in the calculations – quantum field theories use a technique called renormalization to handle this issue, but this cannot be applied to gravity.

Since the 1970s, a number of possible TOEs have been put forward, though all have their problems. They include string and superstring theories (see page 358), and 11-dimensional M-theory (see page 361). A model known as loop quantum gravity, which predicted that space and time themselves are 'quantized' into tiny indivisible units, appears to have been disproven by recent observations from orbiting telescopes.

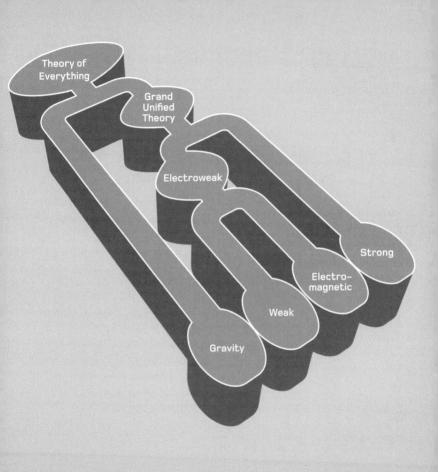

String theories

String theory is a potential 'Theory of Everything' that hopes to explain the four fundamental forces and the origins of the elementary particles. The basic idea is fairly easy to grasp. Objects that appear to be particles are in fact tiny vibrating closed loops or 'strings' of energy, subject to a 'tension' that is related to their size. These loops oscillate at various harmonic frequencies (see page 76), and these frequencies determine the specific quantum numbers that it can adopt (see page 306).

The simplest form of string theory, known as bosonic string theory, was developed in the 1960s. However, it soon revealed a number of problems – for instance, it can only produce bosons and not fermions, it predicts particles with 'imaginary' mass and it can only work mathematically if strings exist in no less than 26 dimensions (see page 360). Supersymmetry (see page 352) offers a possible explanation for fermions and reduces the number of required dimensions to ten. Models that incorporate this idea are called superstring theories.

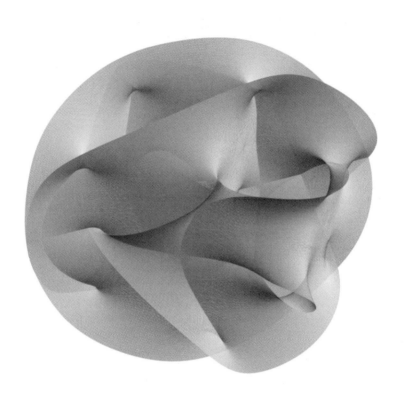

Extra dimensions

According to our immediate perceptions, the large-scale Universe contains four dimensions – three of space and one of time. Combined together, they form a 'spacetime manifold' that can be used to describe everything from Newtonian gravity to general relativity. But since the 1960s, in the search for a Theory of Everything to unite the four fundamental forces of the Universe, theoretical physicists have increasingly fallen back on the idea of unseen additional dimensions. The earliest forms of 'bosonic' string theory called for a total of 26 dimensions, while more advanced 'superstring' theories reduce this number to ten.

In both cases, extra dimensons are required in order to make the mathematical results conform to other properties of the Universe. However, in order to reconcile them with everyday four-dimensional spacetime, the extra dimensions must be explained away somehow. One suggestion is that they are 'compactified' – tightly curled up, wrapped around each other or otherwise reduced in size to a scale far below our ability to measure or detect. Another suggestion, called brane theory, suggests that our four-dimensional Universe is a mere subset, or 'D-brane', afloat in a multidimensional space and the other dimensions lie beyond our perception. According to this idea, gravity 'leaks out' across the higher dimensions, explaining why it is so much weaker than the other fundamental forces. Occasional collisions between narrowly spaced D-branes (on a trillion-year timescale) have also been proposed as a possible mechanism for triggering Big-Bang-scale events to create new Universes (see page 398).

In fact, brane theory usually involves not ten but *11* dimensions. It developed in the late 1990s from an extended string theory called M-theory, which itself adds yet another dimension in an attempt to reconcile five rival ten-dimensional superstring theories. Just as at least three of the four fundamental forces appear to be aspects of a single unifying 'superforce', so the addition of another dimension allows the rival superstring models to be treated as reflections of a single unified M-theory.

Large Hadron Collider

Located up to 175 metres (575 ft) beneath the Swiss–French border, the Large Hadron Collider (LHC) is the world's most powerful particle accelerator (see page 326) and the biggest scientific project in history. Completed by CERN, the European Organization for Nuclear Research, in 2008, it consists of a circular tunnel 27 kilometres (17 miles) in circumference, lined with superconducting electromagnets and interrupted by huge chambers holding experiments and detectors.

The LHC collides beams of protons or lead nuclei to release bursts of energy with intensities not seen since the Big Bang, equivalent to temperatures of many trillions of degrees. In the aftermath of these collisions, the energy rapidly converts back into showers of particles in which short-lived massive particles are briefly stable and available to study. Perhaps the LHC's greatest achievement so far has been the detection of a new massive particle that matches predictions for the Higgs boson (see page 348), potentially confirming the Standard Model.

Relativity and cosmology

Cosmology is physics on a grand scale – the investigation of the origins, fate and structure of the Universe as a whole. While cosmological theories must utimately dovetail with other branches of astronomy to explain how and why individual objects such as galaxies, stars and planets form, in practice cosmologists also concern themselves with explaining the way that features of the Universe arose and how these are linked to the fundamental physics of forces and particles.

Just one of the fundamental forces, gravity, dominates and shapes the structure of the Universe. Although it is feeble compared to electromagnetic and weak and strong nuclear forces, and only makes its presence felt when matter accumulates in substantial masses, it is the only force that can have effects across all distance scales.

In most situations, the effects of gravity and other mechanical interactions can be accurately described by classical laws, such as Newton's law of universal gravitation (see page 32). However, early in the 20th century Albert Einstein outlined a more complete theory of mechanics in the form of special relativity, describing the strange effects that can arise when events in one 'frame of reference' are seen from another one that is in relative motion.

Understanding special relativity is important to the way we interpret observations about the Universe as a whole, where situations involving motion at relatively high speeds are commonplace. Einstein's work on the special theory led him to develop the even wider-ranging theory of general relativity, which revolutionizes the way we look at space and time, and helps describe strange phenomena such as black holes and wormholes.

But while modern cosmology does a good job of describing many aspects of the Universe around us, there are still some important unanswered questions. Was there anything here before the Big Bang that created the Universe 13.8 billion years ago? Why does our Universe display this precise mix of energy and matter? Are there other Universes? And perhaps most soberingly of all, will the Universe ever come to an end?

The speed of light

The theory of relativity can trace its origins back to a single remarkable observation — the fact that in a vacuum, light always travels at the same speed (299,792 kilometres per second or 186,282 miles per second, denoted 'c') regardless of relative motion. In other words, even if you were moving towards one light source and away from another at half the speed of light (0.5 c), light would still reach you from both at the same speed (c). The enormous speed of light and limited speeds of everyday motion meant this only became clear once c could be measured with great accuracy, from around 1860.

In 1887, US physicists Albert Michelson and Edward Morley devised an experiment that should have detected variations in c caused by Earth's motion through a theorized 'luminiferous aether' that was thought to act as a medium for light waves. The experiment's failure not only disproved the aether, but also paved the way for both quantum physics (through the concept of the photon, see page 142), and special relativity.

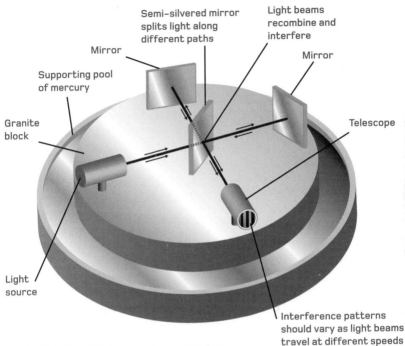

Mirror

Semi-silvered mirror splits light along different paths

Light beams recombine and interfere

Supporting pool of mercury

Mirror

Granite block

Telescope

Light source

Interference patterns should vary as light beams travel at different speeds through the aether

The Michelson-Morley experiment aimed to detect variations in the speed of light travelling along different paths due to Earth's motion through the 'luminiferous aether'.

Special relativity

In a landmark paper published in 1905, Albert Einstein set out a remarkable new view of the Universe. The fixed speed of light, he argued, robs us of any fixed 'frame of reference' with which to measure the physical laws of the Universe. This was, in effect, a restatement of the 'principle of relativity' identified by Italian scientist Galileo Galilei around 1630, which had been undermined in the 1800s by the idea of a 'luminiferous aether' providing an absolute frame of reference (see page 366).

Einstein developed his theory based on two postulates: that the speed of light in vacuum is constant and that the principle of relativity applies between 'inertial reference frames' (those in the special situation of uniform motion with no acceleration – hence 'special relativity'). He showed how in most situations, these assumptions would normally give rise to the familiar laws of classical physics, but in situations involving motion at 'relativistic' speeds approaching the speed of light, they would give rise to some very unusual effects (see pages 370–7).

Time dilation and the Lorentz factor

In essence, time dilation is a slowing-down in the passage of time for objects in one frame of reference compared to those in another, when the two are in uniform motion relative to each other at relativistic speeds (approaching c). A common example is an observer on Earth viewing a distant spacecraft travelling at an appreciable fraction of the speed of light: the observer will see time appearing to pass more slowly for the crew of the spaceship (and conversely, the crew of the spaceship will see time passing more slowly on Earth) – neither frame of reference is preferred provided both are inertial (neither is accelerating with respect to the other).

There are various ways of describing the origins of time dilation, but perhaps the simplest involves considering Einstein's basic postulates: if a measurable reference beam of light is fired at the spaceship, then both the observer and the crew must be able to measure it as having the same speed c. The only way this can be achieved is for the crew of spaceship to experience time more slowly, so that from the point of view of the external observer, they are taking longer to measure the beam's speed.

Although there are many alternatives explanations, this strange phenomenon is no illusion – it is objectively real, and can be best described as a change or warping of 'spacetime' around both objects when viewed from the opposing frame of reference. What's more, time dilation has been proven by several experiments – typically using synchronized atomic clocks (see page 182), one of which is kept in a laboratory while the other is flown at high speed on an aircraft or spacecraft.

When treated mathematically, the dilated time $\Delta t'$ equivalent to time Δt in a frame of reference moving at relative velocity v is given by:

$$\Delta t' = \Delta t \times \frac{1}{\sqrt{(1-v^2/c^2)}}$$

The factor $1/\sqrt{(1-v^2/c^2)}$ crops up throughout the equations of relativity and elsewhere (see, for instance, page 374). It is known as the Lorentz factor (after Dutch physicist Hendrik Lorentz) and denoted by the Greek letter γ (gamma). Hence, the equation above can be restated as: $\Delta t' = \gamma \Delta t$.

Lorentz contraction

According to special relativity, when an object reaches relativistic speeds compared to an observer, it undergoes a contraction in length along its direction of travel. This contraction was independently proposed by Irish physicist George Fitzgerald and Dutch scientist Hendrik Lorentz around 1890 as a possible cause of the negative Michelson–Morley result (see page 366). It is commonly known as the Lorentz contraction, but it was only properly explained by Einstein as a result of relative motion.

The contraction is often explained as an effect of light leaving different parts of the object at different times in order to reach an observer simultaneously. This might seem to suggest that it is an optical illusion, but in fact the effect is as objectively real as any other aspect of relativity: it arises from the same distortions in the perception of spacetime that create the time dilation effect, and unsurprisingly, the equations that describe it also make use of the Lorentz factor (see page 370).

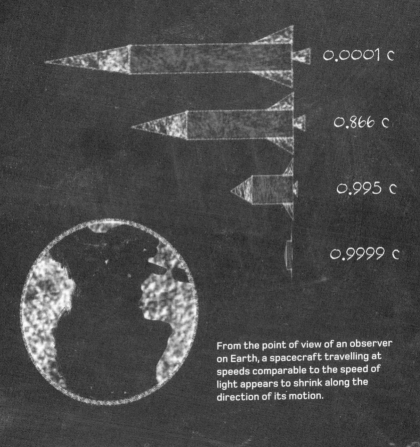

0.0001 c

0.866 c

0.995 c

0.9999 c

From the point of view of an observer on Earth, a spacecraft travelling at speeds comparable to the speed of light appears to shrink along the direction of its motion.

Mass–energy equivalence

The universal speed limit imposed by special relativity poses an interesting problem – how can it be reconciled with the conservation of energy and momentum? In a 1905 paper, Einstein explained the solution: at speeds approaching the speed of light, energy supplied to an object is used to increase the object's mass as well as its velocity. This allows its energy and momentum to increase while limiting further acceleration.

Mathematically, the energy E, mass m and momentum p of an object moving at speed v are given by the simple equations:

$$E = \gamma E_0 \ , \quad m = \gamma m_0 \quad \text{and} \quad p = \gamma m v$$

where γ is the Lorentz factor (see page 370), E_0 is the object's 'rest energy' when it is stationary relative to the observer's frame of reference, and m_0 is its 'rest mass'. Based on this relativistic 'dilation' of mass, Einstein was able to show that mass and energy are actually equivalent in all situations, giving rise to the famous equation $E = mc^2$.

$$E=mc^2$$

The twin paradox

One of the most famous paradoxes in relativity describes the different experiences of time for a twin who stays on Earth and one who goes on a spaceflight at relativistic speeds and then returns home. Thanks to the effects of time dilation (see page 370), the space-travelling twin will not have aged as much as the one that stayed at home – a result that has been verified on a small scale by atomic clocks aboard spacecraft.

The paradox arises when we ask how the Universe remains consistent – surely *each* twin should see their sibling's life slowed down by time dilation, so how does the age difference arise? In 1913, German physicist Max von Laue pointed out that the similarity is misleading – the twins' experiences differ because one has undergone acceleration and deceleration, and so has not remained in the same inertial reference frame throughout. Einstein and Max Born later explained the differences through 'gravitational time dilation' experienced during acceleration and deceleration (see page 380).

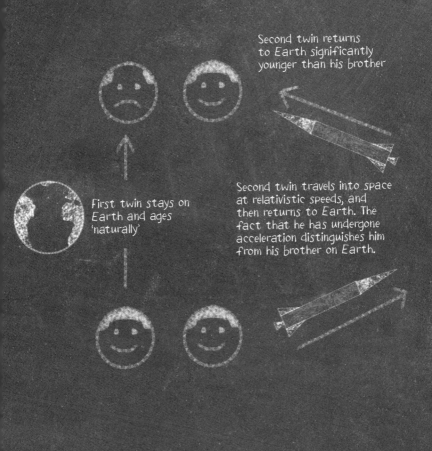

Second twin returns to Earth significantly younger than his brother

First twin stays on Earth and ages 'naturally'

Second twin travels into space at relativistic speeds, and then returns to Earth. The fact that he has undergone acceleration distinguishes him from his brother on Earth.

Spacetime

Key to the concepts of relativity is the idea that space and time are interchangeable – an approach developed by Einstein's former tutor, German physicist Hermann Minkowski, in the years after the 1905 publication of special relativity. Minkowski's 'spacetime' unites three 'space-like' and one 'time-like' dimensions in a single four-dimensional 'manifold', and treats many of the effects of special relativity as 'rotations' of coordinates within this spacetime. Such transformations have the effect of shortening space-like dimensions (the Lorentz contraction) or lengthening the time-like one (time dilation).

The idea that spacetime can be distorted by the presence of mass lies at the heart of general relativity (see page 380), and therefore explains how gravity can give rise to many of the same effects as relativistic motion. In this context, spacetime is often visualized by ignoring one of the space dimensions completely and considering space as a 'rubber sheet' with indentations indicating large masses and gravitational fields.

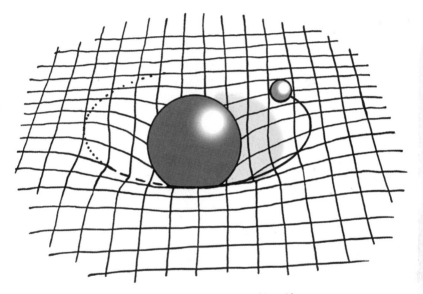

Minkowski's concept of spacetime combines three space dimensions with one of time. Large masses distort spacetime like heavy weights on a rubber sheet, and other objects follow orbits around these gravitational 'wells'.

General relativity

In 1915, Einstein published a theory of relativity that showed how the laws of physics work when comparing frames of reference that are *not* necessarily inertial (in other words, incorporating those which are experiencing acceleration as well as those in the 'special' situation of uniform motion). Einstein's new theory of 'general relativity' was inspired by his realization, in 1907, that from the point of view of physical laws, an observer in a gravitational field should experience the same effects as one undergoing constant acceleration.

Coupled with Minkowski's ideas about spacetime (see page 378) this inspired Einstein to develop a model of gravitational fields as distortions of time and space creating effects that were similar to those arising in special relativity. An early success for the theory came when Einstein used it to explain hitherto inexplicable behaviour of Mercury's orbit, but general relativity was only widely accepted following the first observations of gravitational lensing in 1919 (see page 382).

Measurement on Earth

Measurement in rocket

An observer in a stationary box on the surface of the Earth can conduct an experiment to measure the force of gravity. However, his situation is indistinguishable from what he might measure on board a steadily accelerating rocket. This is the 'principle of equivalence' that lies behind general relativity.

Gravitational lensing

One of the most useful predictions of general relativity is that gravitational fields deflect the passage of light rays that pass nearby. This effect, known as gravitational lensing, cannot be fully explained by classical gravity (since photons are massless and should therefore be immune to Newton's law of universal gravitation), but is a natural consequence of 'warped' spacetime around massive objects such as stars. Detecting gravitational lensing therefore became a key test for general relativity, and in 1919 British astronomer Arthur Eddington led an expedition to the Atlantic island of Principe, where he successfully measured shifts in the apparent locations of stars around the Sun during a total solar eclipse.

Today, lensing is more than just a curiosity and a useful confirmation of Einstein's theory – astronomers can use the strength of lensing distortions to measure the mass of distant galaxy clusters, and can even use its magnifying effect to track down galaxies at the very limits of the visible Universe.

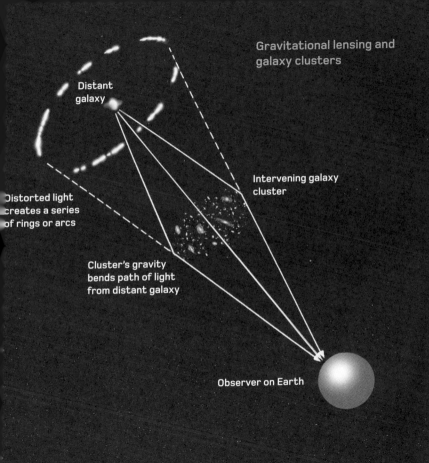

Gravitational lensing and galaxy clusters

Distant galaxy

Intervening galaxy cluster

Distorted light creates a series of rings or arcs

Cluster's gravity bends path of light from distant galaxy

Observer on Earth

Gravitational waves

One of the major unconfirmed predictions of general relativity is the existence of gravitational waves – disturbances in spacetime that spread out in ripples around massive, asymmetrical objects. Potential sources of such waves include binary systems in which dense objects such as white dwarfs, neutron stars or black holes slowly spiral in towards each other. Because they diminish only very gradually as they travel across space, and their effects are not blocked by intervening material, gravitational waves should be detectable over vast distances.

The passage of a wave past Earth would create a distinctive distortion in the dimensions of space that physicists hope to detect by using interferometers to measure tiny changes in the distance travelled by laser beams (see page 102). So far, Earth-based instruments have been too insensitive to measure such effects, but there are sound hopes that they will be detected in the near future by space-based interferometers.

This artist's impression depicts the gravitational waves generated by a pair of black holes in orbit around one another.

Singularities
and black holes

A singularity is a point in spacetime at which the laws of physics break down and the equations that model them produce nonsensical results. They tend to be enormous masses concentrated in a tiny point in space, whose powerful gravity seals them off from the Universe. This creates a region whose 'escape velocity' – the speed at which an object has to move to escape its gravity – exceeds the speed of light: a black hole. Singularities can exist on any scale, but the ones so far detected are formed by the sudden collapse of giant stellar cores in supernova explosions or, on a far larger 'supermassive' scale, in the crowded central regions of galaxies.

The closest visible surface to a black hole, known as its event horizon, emits a form of radiation called Hawking radiation (see page 394). However, black holes are usually detected either from their gravitational effects or by the fierce radiation emitted as matter pulled into the black hole's grasp is heated to millions of degrees by the extreme gravity.

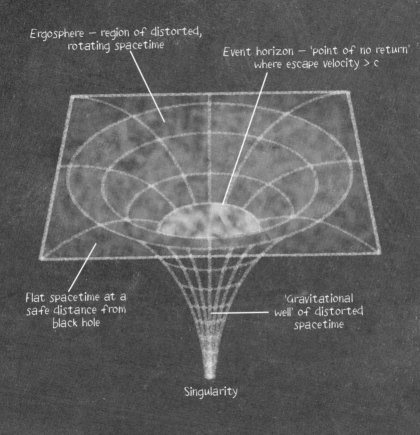

Ergosphere — region of distorted, rotating spacetime

Event horizon — 'point of no return' where escape velocity > c

Flat spacetime at a safe distance from black hole

'Gravitational well' of distorted spacetime

Singularity

Wormholes

In 1935, Albert Einstein and Nathan Rosen published a new study of general relativity, analysing the 'field equations' that describe spacetime and producing a solution in which the formation of a black hole would naturally create a 'white hole' elsewhere in the Universe. This hypothetical black/white hole pair would be connected by a tunnel of spacetime called an Einstein-Rosen bridge, or more commonly a wormhole. If, as astronomical evidence suggests, spacetime is 'curved' on the largest scales, then the distance between different parts of the Universe may be much shorter through a wormhole than across normal space.

Travel through a natural wormhole is rendered impossible by the singularity at the centre of the bridge. However in 1988, Caltech physicists Kip Thorne and Mike Morris showed that it might be possible to create a traversable wormhole without a central singularity, perhaps using 'exotic' matter that takes advantage of the Casimir effect (see page 340).

Anatomy of a wormhole

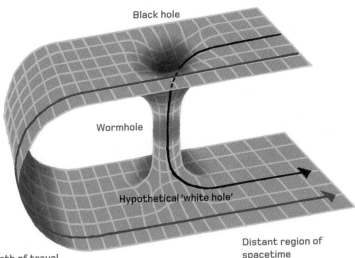

Nearby region of spacetime

Black hole

Wormhole

Hypothetical 'white hole'

Distant region of spacetime

Path of travel through normal space

Time machines

Based on their initial work on traversable wormholes (see page 388), Kip Thorne, Mike Morris and their colleague Ulvi Yurtsever showed in 1988 how these shortcuts across spacetime could theoretically be used to build a time machine. The principle relies on accelerating one end of the wormhole to relativistic speeds so that it experiences significant time dilation and then bringing the two ends of the wormhole (now connecting different times) together.

This sort of wormhole time machine is effectively a tunnel through time but, significantly, would not allow travel back to before the machine was created, helping to avoid paradoxes (see page 392). Time dilation can also function in the framework of relativity as a sort of time travel, although it only offers a one-way ticket to the future (see page 376). However, while faster-than-light (FTL) travel, if it were possible, is predicted to lead to backward time travel, this is somewhat moot since 'genuine' FTL motion is forbidden by relativity.

The grandfather paradox

The hypothetical possibility of 'backward' time travel gives rise to important questions about causality, since in theory a traveller journeying into their own past could violate the principle that causes precede effects. The classic time-travel paradox was first explored by French writer René Barjavel in a novel of 1943: if you travel back in time and kill your own grandfather before your father is conceived, then how can you have existed to travel back in time in the first place?

Solutions to such issues include the Novikov self-consistency principle (in which the only permitted 'timelines' are ones that are consistent, and paradoxes are not permitted to arise); the Huggins displacement theory (in which time travel is accompanied by a displacement in space that prevents breaches of causality); and the creation of parallel Universes, entertainingly explored in the *Back to the Future* movies. Another solution, of course, is that backward time travel is simply impossible.

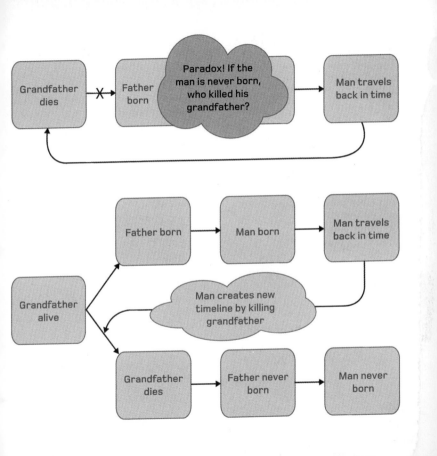

Hawking radiation

According to relativity alone, a black hole should be entirely black, with no radiation of any kind able to escape from its event horizon. As such, it should only ever grow in mass, since it cannot lose energy. However, in 1974 Stephen Hawking showed how quantum physics and relativity could combine to produce distinctive emission at the event horizon – Hawking radiation.

The radiation is created by quantum fluctuations that produce 'virtual' particle-antiparticle pairs (see page 338). Normally such pairs collide and 'annihilate' almost instantaneously, releasing their energy back into the vacuum, but if a pair forms exactly on the event horizon, one particle may be pulled into the black hole while the other escapes to become 'real', draining energy from the black hole in the process. Seen from a distance, the effect is to make the black hole emit very low-temperature black body radiation (see page 148), losing mass over very long time periods (many times the age of the Universe). However, such radiation has not yet been conclusively detected.

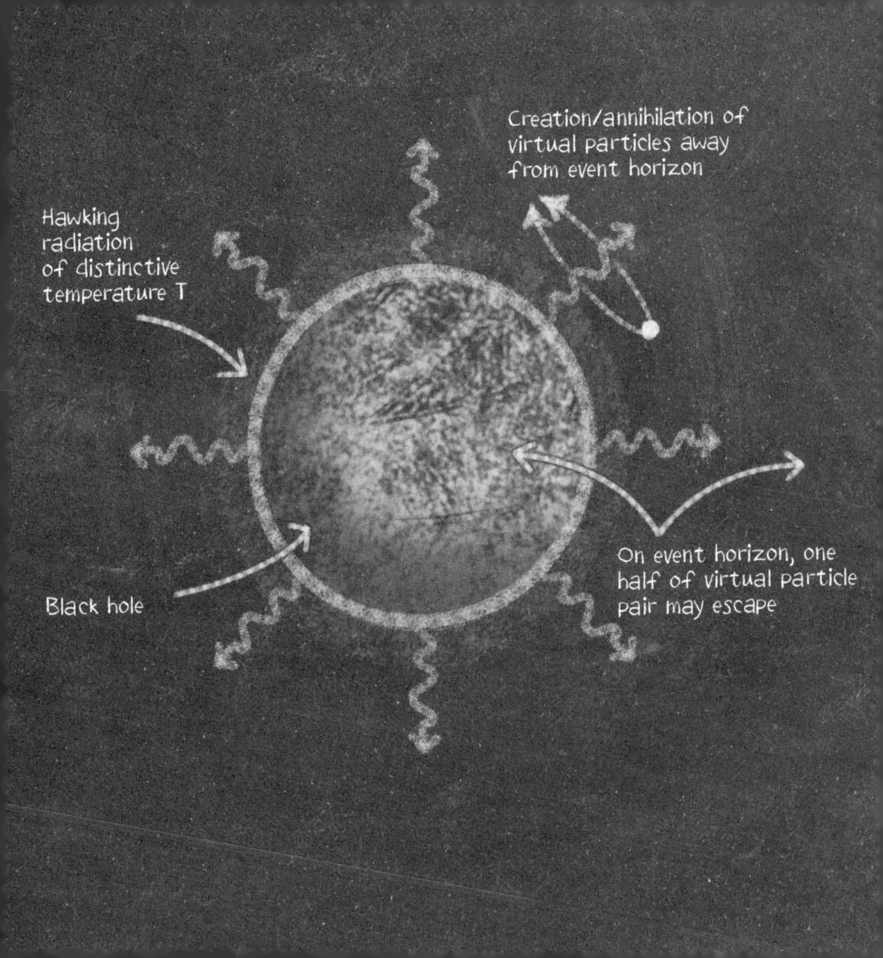

Creation/annihilation of
virtual particles away
from event horizon

Hawking
radiation
of distinctive
temperature T

On event horizon, one
half of virtual particle
pair may escape

Black hole

Hubble's law

In the early 1900s, astronomical evidence suggested that the Universe was relatively small (perhaps 100,000 light years across), packed with matter and unchanging in size. This raised the obvious question of why it does not collapse under its own gravity and forced Einstein to introduce an anti-gravitational force, the 'cosmological constant', into general relativity.

Within a few years, however, a revolution in astronomy showed that in this case at least, Einstein was wrong. In the mid-1920s, US astronomer Edwin Hubble proved that remote clouds of stars were independent galaxies millions of light years away. What was more, by identifying Doppler-shifted spectral lines in the light of these distant galaxies (see page 72), he discovered that the further away a galaxy lies from Earth, the faster it is receding. This relationship, known as Hubble's law, shows that the Universe as a whole is expanding, pulling matter apart unless local gravity is strong enough to resist. It naturally gives rise to the idea of a 'Big Bang' (see page 398).

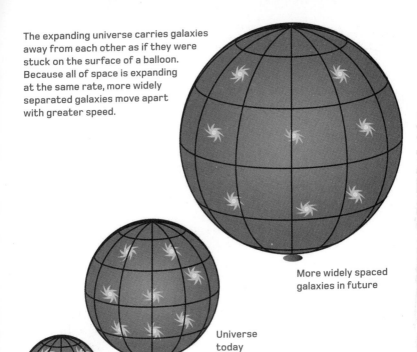

The expanding universe carries galaxies away from each other as if they were stuck on the surface of a balloon. Because all of space is expanding at the same rate, more widely separated galaxies move apart with greater speed.

More widely spaced galaxies in future

Universe today

Galaxies closely packed in distant past

The Big Bang

The discovery that the Universe is expanding (see page 396) naturally suggests that it was much smaller in the distant past. Belgian physicist Abbé Georges Lemaître was the first person to trace cosmic expansion back to its logical starting point, suggesting in 1931 that the Universe began life in an extremely dense, very hot state he called the 'primeval atom'. Established physicists of the time, however, were uncomfortable with the idea of a 'beginning' to the Universe, and put forward a variety of other theories to reconcile expansion with an eternal and unchanging Universe.

It was only in the late 1940s, following advances in nuclear physics, that physicists Ralph Alpher and George Gamow produced a theory of how the hot conditions of the primeval atom could produce the range of light elements (predominantly hydrogen and helium) found in the raw material of the Universe. British astronomer Fred Hoyle, a proponent of the rival 'steady state' theory, derided the idea as nothing but a 'big bang', but

Alpher and Gamow's description of the Universe was backed up in 1964 by the discovery of the Cosmic Microwave Background Radiation – a faint glow from the distant Universe whose features cannot be adquately explained by any other theory (see page 400).

In its current form, the Big Bang theory describes how the Universe was born in a spontaneous explosion 13.8 billion years ago. Conditions in the initial fireball were so intense that the four fundamental forces were unified for a brief instant, before their separation drove a dramatic 'growth spurt' called Inflation. In the following three minutes, as temperatures fell away, particles 'condensed' out of a seething sea of energy – quarks at first, and then much lighter electrons. The quarks bonded to form protons and neutrons, and these in turn rapidly bound together in light atomic nuclei.

After three minutes, temperatures fell to a point where matter could no longer be created. The Universe remained an expanding opaque fireball with photons of light ricocheting through a dense fog of particles until, after 380,000 years, temperatures fell to a point where electrons could bind together with nuclei to form the first atoms.

Cosmic Microwave Background Radiation

Discovered in 1964, the Cosmic Microwave Background Radiation (CMBR) is a faint microwave signal from all parts of the sky, corresponding to a temperature just 2.7 kelvin above absolute zero (see page 108). It is the afterglow of creation, released as the Universe became transparent 380,000 years after the Big Bang. Since its discovery, the CMBR has been subjected to intensive study, particularly by satellite observatories such as COBE, WMAP and Planck. They have detected 'ripples' of just a few millionths of a kelvin that are believed to have formed from a combination of random quantum-level fluctuations present in the Big Bang and magnified by Inflation (see page 399), and 'acoustic oscillations' caused by shockwaves rearranging matter in the subsequent expanding fireball. The ripples are generally believed to be the seeds for large-scale structure in the later Universe, such as galaxy clusters and superclusters, and can also reveal information about cosmic features such as 'dark matter' and 'dark energy' (see pages 402 and 404).

A map of the CMBR from NASA's Wilkinson Microwave Anisotropy Probe (WMAP) reveals tiny fluctuations between hotter areas of the primordial Universe (bright spots) and cooler ones (dark regions).

Dark matter

Several lines of astronomical evidence suggest a remarkable fact – an estimated 84.5 per cent of all the mass in the Universe is tied up in invisible 'dark matter', with visible matter of the observable Universe accounting for just a small fraction of the cosmos. Dark matter gives away its presence through its gravity altering the rotation of spiral galaxies and the motion of individual members of galaxy clusters.

While small amounts of dark matter could be accounted for by ordinary matter in invisible forms (such as black holes), most scientists agree that the solution to this mystery probably lies with a new type of 'Weakly Interactive Massive Particle' (WIMP). Such particles could be cold and relatively slow moving, and are more likely to be discovered in particle accelerator experiments than through astronomical studies. Since its discovery in the 1930s, the existence and amount of dark matter were thought to control the Universe's ultimate fate, but the discovery of 'dark energy' (see page 404) complicates matters.

A computer simulation maps the web of dark matter surrounding visible galaxies in the Universe.

Dark energy

During the late 1990s, astronomers studying supernovae in remote galaxies made a remarkable discovery – at the greatest distances, these enormous stellar explosions appear systematically fainter than their Doppler shifts suggest they should be (see page 396). The widely agreed explanation is that something is driving the Universe's expansion to speed up – contrary to the long-held assumption that the expansion should be slowing down in the aftermath of the Big Bang.

The driving force behind this acceleration has become known as 'dark energy' and appears to make up some 68 per cent of all the energy content of the Universe (with normal and dark matter accounting for the rest). Physicists are still arguing over its origin and whether it is related to zero-point energy (see page 340) and/or the 'cosmological constant' in Einstein's original theory of general relativity (see page 380). But whatever its origin, dark energy seems to confirm that the Universe will keep expanding forever, overcoming the inward pull of gravity.

What is the Universe made of?

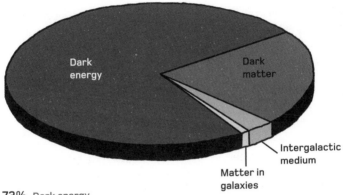

Dark energy

Dark matter

Intergalactic medium

Matter in galaxies

73% Dark energy

23% Dark matter

3.6% Ordinary atoms in intergalactic medium

0.4% Ordinary atoms within galaxies

The anthropic principle

Perhaps the single strangest aspect of our Universe is simply the fact that intelligent human beings are here to see and experience it. Woven throughout physics are numerous constants – for instance, the precise strengths of fundamental forces – that would result in a radically different Universe if their values were even slightly different. So are the laws of physics 'made for life'? Not necessarily, it seems.
"

In 1973, Australian physicist Brandon Carter proposed an 'anthropic' approach to cosmology – one that takes into account the presence of humans within the Universe. At the simplest level, the implications of this principle are quite straightforward: we should not be too surprised at some of the apparent 'fine-tuning' we find in the laws of physics, since if the various constants did not have their observed values, we probably wouldn't be around to observe them. Carter's approach has subsequently become known as the 'weak anthropic principle', in contrast to a 'strong principle', which suggests that, for one reason or another, the Universe *has* to be fine-tuned for life.

Both versions of the prinicple raise certain questions – if the Universe is a one-off, then surely the odds against it randomly producing such hospitable conditions are astronomical? Or is our Universe merely one of many? Possible solutions to these questions include:

- The 'new physics' theory: the fine-tuning of the Universe is shaped by some as-yet-undiscovered physical laws.
- The multiverse theory: our Universe is one of an endless succession, either successive in time or parallel in higher dimensions, in which all configurations are eventually explored.
- The participatory theory: some interpretations of quantum physics require there to be a conscious observer – in other words, the Universe couldn't exist without us.
- The design theory: the Universe was tailor-made for life by an external designer (be that a religious deity or alien intelligence).

Whether the answer lies with any of these approaches or with something else entirely, we can be sure physicists will be at the forefront of future discoveries.

Glossary

Alpha particle
A particle released by radioactive decay that consists of two protons and two neutrons – equivalent to the nucleus of a helium atom.

Angular momentum
A property of rotating objects analogous to momentum, and linked to their inertia and rate of rotation around an axis of rotation.

Atomic mass
A measure of the mass of any atom in 'atomic mass units' – in practice, equivalent to the total number of protons plus neutrons it contains.

Atomic number
A property that indicates the number of protons in an atomic nucleus, and therefore the number of electrons in the corresponding neutral atom. An atom's atomic number defines which element it forms.

Beta particle
A particle released by radioactive beta decay – usually an electron, but rarely a positron. Beta particles are released from unstable atomic nuclei when a neutron transforms into a proton or, more rarely, vice versa.

Boson
A particle with a zero or whole-number 'spin'. Bosons display unique physics because they are unaffected by the Pauli exclusion principle.

Conventional current
The notional direction in which positive charge flows around a circuit – opposite to the physical flow of electrons.

Cosmic rays
High-energy particles generated by various processes in deep space, which give rise to showers of lower-energy particles as they interact with Earth's upper atmosphere.

Electromagnetic radiation

A natural wave phenomenon consisting of electrical and magnetic waves interfering with and reinforcing each other. It can exhibit very different properties depending on its wavelength, frequency and energy.

Electron

A low-mass elementary particle carrying negative electrical charge. Electrons are found in the orbital shells surrounding an atomic nucleus, and play a key role in chemical bonding. They are also the principal means by which electric current flows through materials.

Endothermic process

Any chemical or physical process that absorbs energy from the surrounding environment.

Exothermic process

Any chemical or physical process that generates and releases an excess of energy.

Exponential decay

A common pattern found in physical processes in which the rate of a process decreases in proportion to its value. For example, radioactive decay, in which the rate of decay is proportional to the number of radioactive nuclei present in a material, and therefore decreases over time.

Fermion

Any particle with a half-integer 'spin' property, including all the elementary matter particles (known as quarks and leptons). Fermions are governed by the Pauli exclusion principle, which limits their behaviour and explains much of the structure of matter.

Frame of reference

Any consistent system of coordinates that can be used to measure the properties of objects. Frames of reference in motion to one another are particularly relevant to the study of relativity, since observations made in different reference frames can differ greatly.

Gamma radiation
A form of high-energy electromagnetic radiation released by various processes such as radioactive decay.

Inverse square law
A pattern found in physical processes in which the strength of a property such as a force decreases as the square of the distance from its source increases, as the force becomes more 'spread out' across a spherical region of space.

Isotope
A term used to distinguish between atoms of the same elements with different masses, due to differing numbers of neutrons in their atomic nuclei.

Lepton
Any member of a family of elementary particles that are not susceptible to the strong nuclear force, including electrons and neutrinos.

Magnetic moment
A property determining the strength of the magnetic field created by an object, and its susceptibility to the influence of other magnetic fields.

Momentum
A property found by multiplying an object's mass by its velocity in a particular direction. Momentum determines the force required to stop an object, or accelerate it to a particular velocity.

Neutron
An electrically neutral subatomic particle with substantial mass, found in the nuclei of atoms.

Nucleon
A catch-all term for the principal particles found in atomic nuclei – protons and neutrons. The number of nucleons in an atom determines its atomic mass.

Orbital shell
A region surrounding an atomic nucleus, in which electrons are found.

Pauli exclusion principle
A law that prevents fermion particles from occupying identical 'states' in a system, and is therefore responsible for much of the structure of matter.

Photon
A discrete burst or 'packet' of electromagnetic energy that can display wave-like, as well as particle-like, behaviour.

Planck's constant
A physical constant that helps define quantum-scale relations such as that between the frequency of a photon and the energy it contains.

Potential difference
A measure of the 'electric potential energy' between two points. Measured in volts, it indicates the work that must be done in moving a unit of electric charge through the electric field between the points.

Proton
A subatomic particle with substantial mass and positive electric charge, normally found in the atomic nucleus.

Quantum
The minimum possible amount of a particular physical property that may be involved in a physical interaction. Certain phenomena, such as the energy of light waves and of electrons in an atom, are inherently 'quantized' on the smallest scale. By extension, quantum physics is a term used to describe the strange and sometimes counterintuitive physics that occurs on very small, subatomic scales.

Radioisotope
An isotope of an atom that is unstable (typically due to an excess of neutrons over protons in its nucleus) and prone to undergoing radioactive decay in order to reach a more stable configuration.

Spin
A property of subatomic particles, analogous to angular momentum in larger objects, which affects many aspects of their behaviour.

Vacuum tube
An electronic device that controls the flow of electric current from a beam fired through a vacuum between two electrodes.

Index

Quercus Editions Ltd
55 Baker Street
7th Floor, South Block
London
W1U 8EW

First published in 2014

Copyright © Quercus Editions Ltd 2014
Text by Giles Sparrow
Design and editorial: Pikaia imaging
Design assistant: Kathryn Brown

A catalogue record of this book is available from the British
Library

ISBN 978 1 78206 648 4

Printed and bound in China

10 9 8 7 6 5 4 3 2 1

Picture credits 2: Maximilien Brice,
CERN; 17: Shutterstock/dslaven;
59: Wikimol; 67: Shutterstock/
elen_studio; 79: Shutterstock/
Mopic; 85: Shutterstock/
rangizzz; 129: Shutterstock/
James Doss; 151: United States
Nuclear Regulatory Commission;
237: Shutterstock/Felix-
Andrei Constantinescu; 249:
Shutterstock/Alex Mit; 267: Tomasz
Barszczak/Super-Kamiokande
Collaboration/Science Photo
Library; 293: US Geological Survey;
313: Shutterstock/Thomas Lenne;
321: NIST/JILA/CU-Boulder; 335:
Shutterstock/Ivancovlad; 349:
Lucas Taylor/CERN; 351: Claudio
Rocchini; 359: Lunch @ en.wikipedia;
363: Maximilien Brice, CERN; 385:
T. Carnahan (NASA GSFC); 391:
Shutterstock/Daniel Boom; 401:
WMAP Science Team, NASA; 403:
Springel et al. (2005).

All other illustrations by Tim Brown,
except 4-5, 37, 39, 62-3, 123, 126-7,
161, 189, 195, 221, 319 and 379 by
Patrick Nugent, Guy Harvey and
Nathan Martin